AF255589

# The Performance Review Handbook.

*How New Managers Write Better Reviews, Deliver Better Feedback, and Build Stronger Teams*

By

**Kevin E Baker**

Copyright © **2026 Kevin Baker**

All rights reserved.

No part of this publication may be reproduced, distributed, or transmitted in any form or by any means, including photocopying, recording, or other electronic or mechanical methods, without the prior written permission of the publisher, except in the case of brief quotations embodied in critical reviews and certain other non commercial uses permitted by copyright law.

PUBLISHED

BY

**Kevin Baker**

# Table of Contents

**Kevin Baker**

**Kevin Baker**

# Dedication

*For my beautiful wife Monica- Thank you for making me a better man.*

*And my AHR family. It was truly an honor.*

**Kevin Baker**

# Introduction

# The Review Nobody Taught You How to Write

Nobody teaches you how to write a performance review.

You are promoted into management, or hired into it, or handed a team and a title and told to figure it out, and at some point in your first December, a deadline appears in your calendar that says PERFORMANCE REVIEWS DUE and you open a blank document and realize that no one, at any point in your professional development, told you how to do this.

So you do what most first-time managers do. You write something. You try to be fair. You try to remember what happened in March. You use the phrases you have heard before, "team player," "strong work ethic," "room for growth," because they sound like the right language and because the deadline is tomorrow and you still have seven more to write.

And somewhere in the organization, seven people receive those reviews and read them and feel, with varying degrees of disappointment, that the document in front of them does not really see them. That it could have been written for anyone. That their manager, whatever else might be true about them, was not paying attention.

This book exists because that outcome is preventable. Not just preventable, but correctable, right now, with the next review you write.

The tools in this book come from twenty years of my own performance management experience across a billion-dollar distribution network: from writing reviews for hourly associates on the warehouse floor, to developing the managers who wrote them, to sitting in calibration sessions where ratings were reconciled, language was scrutinized, and the difference between a useful review and a damaging one became visible in real time.

What follows is what was learned in those rooms. The documentation habits, the language frameworks, the legal guardrails, the conversation techniques, the self-review strategies, and the follow-up practices that turn a performance review from an annual compliance exercise into one of the most powerful development tools a manager has.

## Who This Book Is For

This book was written for the first-time manager who has just been handed a performance review form and no instructions. That manager is the primary audience. Every framework, every phrase bank, every worked example in these pages was designed with that person in mind: someone who cares about doing right by their team and simply has not been taught how.

But this book serves a broader audience than that. The experienced manager who has been writing reviews for ten years, on instinct, without a framework, hoping each year to do better, will find something here too. So will the HR professional looking for a training resource that goes beyond policy compliance. And so will the senior leader who wants to raise the standard of performance conversations across an entire organization.

If you manage people who receive annual reviews, this book applies to you. The examples come from distribution operations, specifically from DCs and warehouse floors and logistics networks, because that is where this knowledge was earned. But the frameworks are universal. A review for a software engineer, a nurse, a financial analyst, or a classroom teacher follows the same basic structure: be specific, use evidence, be honest, and follow through. The industry changes. The principles do not.

**A Note on Industry**: The examples in this book come from distribution operations. The frameworks apply everywhere.

The sample reviews in these pages are drawn from distribution center operations: warehouse supervisors, DC managers, regional logistics leads. The metrics are cartons per hour and on-time delivery rates. The teams are hourly workers on overnight shifts. This specificity is intentional. These are real reviews, written for real people, in a real industry, and that specificity is what gives them their power. A generic example teaches you how to sound professional. A real example teaches you how to think.

But the thinking transfers. A nurse manager, a school principal, or a software team lead faces the same fundamental challenges: how to document performance throughout the year, how to apply a rating system fairly, how to give corrective feedback without triggering defensiveness, how to set goals specific enough to be meaningful.

When you encounter an example about cartons per hour or a DC manager navigating a pandemic, translate it to your own context. The industry is the container. The performance management principles are the contents. One is specific to a world you may never work in. The other is yours to take and use.

Read it. Use it. And write the review that gets kept.

FAST START: BREAK GLASS IN EMERGENCY

# The 60-Minute Review

For when the deadline is tomorrow and you haven't started yet.

If your review is due tomorrow, this page is for you. This is not the ideal way to write a performance review, but it is a complete, defensible, and respectful one. Follow the sequence. Don't skip steps. You'll have a real review. Not a great one, but an honest one, in an hour.

## MINUTES 0–10: Pull Your Evidence

Open your email and search the employee's name. Look for moments: a compliment forwarded, a note about a problem, a project milestone. Check your calendar for any one-on-ones or incidents. Write down three to five specific moments from the past year, positive or negative. These are your raw material. If you have a documentation file, open it now instead.

## MINUTES 10–20: Read Their Self-Review (If You Have One)

Don't skim it. Read it once, slowly. Note what they're proud of, what they left out, and whether it matches what you observed. Their own language will tell you something about how they see their work. If you don't have a self-review, skip ahead to the next step.

## MINUTES 20–40: Write Each Section Using O-I-D

For every section of your review form, write one or two sentences using the OID structure. You don't need all three elements in every sentence, but every comment needs all three present somewhere nearby.

O: Observation: What you actually saw or measured. Specific. Dated if possible.

I: Impact: What it produced: for the team, the customer, or the operation.

D: Direction: What you're asking for going forward. Continue it, improve it, or stretch it further.

Example: "Marcus consistently resolved shipping discrepancies within the same shift they were reported, a habit that reduced carrier chargebacks noticeably over Q3. Going into the new year, I'd like to see him bring that same urgency to cross-dock coordination, where delays have occasionally cascaded downstream."

## MINUTES 40–50: Assign Your Rating, Then Check It Against What You Wrote

Pick your rating honestly using your organization's scale. Then read your narrative back and ask yourself: does the writing match the number? If the narrative is positive but the rating is low, one of them needs to change. Fix it before someone else notices the gap.

## MINUTES 50–60: Quick Self-Audit Before You Submit

Read through what you wrote and confirm each of the following:

- Every comment has at least one specific observation. No generalities.
- No protected characteristics mentioned (age, race, health, family situation).
- Rating and narrative tell the same story.
- Nothing here would genuinely surprise the employee. If something would, flag it before the meeting, not during it.

At least one forward-looking statement or development goal is included.

Next year, start earlier. The lasting fix is to start earlier. A simple 15-minute monthly habit eliminates this problem entirely. The managers who never panic at review time aren't doing more work; they're spreading it across the year. This page got you through today. The rest of the book gets you ready for next time.

**Kevin Baker**

# Chapter 1

# The Purpose Nobody Explains to New Managers

*Before you write a single word of any performance review, you need to understand what it actually is, and what it is not. Most first-time managers have never been told this. So they write reviews the way they received them, which is often the wrong way, which is why the cycle of mediocre reviews perpetuates itself from one generation of managers to the next.*

The performance review is one of the most consequential documents a manager produces. It shapes how an employee understands their own work. It influences compensation, promotion, succession planning, and termination decisions. It is read by HR, senior leadership, and sometimes by attorneys. It becomes the permanent official record of someone's professional year.

And most first-time managers approach it as a form to fill out.

That gap between what the performance review actually is and how most managers treat it is where the real damage happens. Not in the rare cases where a manager writes something overtly unfair or legally problematic. In the thousands of reviews written every year by well-meaning managers who did not understand what they were doing, why it mattered, or who was actually affected by it.

This chapter changes that. Before we get to the language, the templates, the frameworks, or the conversation techniques, we start here. With the purpose. Because if you understand what a performance review is actually for, everything that follows will make more sense. And you will write it differently.

## What a Performance Review Actually Is

A performance review is simultaneously four different things. Most managers understand one or two of them. The managers who write the best reviews understand all four and write with awareness of how each one operates.

It Is an Official Record

The moment you submit a performance review through your organization's system, it becomes part of the employee's permanent personnel file. It can be subpoenaed. It can be submitted as evidence in an employment dispute. It can be used to support or challenge a termination, a promotion decision, or a compensation adjustment. Years after the employee has left the organization, years after you have moved to a different role or a different company, that document still exists in the record.

This does not mean you should write reviews that are defensive or lawyerly. It means you should write them with the awareness that your words have weight beyond the conversation you are having today. Write what is true. Write it specifically. Write it in language you would stand behind in any setting.

## It Is a Development Tool

The most useful thing a performance review does (when it is written well) is give an employee a clear, honest, evidence-based picture of where they stand and what the path forward looks like. Not a generic summary of their job. A specific assessment of this person's performance this year, with enough precision that they can do something with it.

A review that tells an employee they are a 'team player with room to grow' tells them nothing. A review that tells them they lead their team effectively day to day but consistently defer to others in cross-functional settings, and that this pattern is limiting their visibility at the senior level, gives them something to act on. Specificity is what makes a review a development tool rather than a formality.

## It Is an Organizational Signal

Every review a manager writes sends a signal to the organization about what is valued, what is expected, and what gets rewarded. When a manager consistently gives high ratings to employees regardless of performance, a pattern sometimes called grade inflation, they signal to the organization that standards do not apply in their team. When a manager consistently writes specific, evidence-based reviews that clearly distinguish between performance levels, they signal that they are a credible evaluator whose assessments can be trusted in succession planning, calibration, and promotion decisions.

The manager whose reviews are trusted is the manager whose people get promoted. The manager whose reviews are not trusted, because they inflate, deflate, or say the same thing about everyone, is the manager whose talent gets overlooked, because no one above them knows how to read what they write.

## It Is a Reflection of the Manager

This is the dimension most first-time managers do not see coming. The review you write is not only an assessment of your employee. It is an assessment of you. The quality of your observation, the precision of your language, the fairness of your standards, the consistency of your documentation: all of it is visible to every person above you who reads what you produced.

A vague review signals a manager who was not watching. An inflated review signals a manager who lacks the courage or judgment to assess accurately. An inconsistent review, where the language does not match the rating, signals a manager who does not understand the system they are operating in. Conversely, a well-written review signals a manager who observes carefully, thinks clearly, communicates precisely, and takes the development of their people seriously. That signal travels further and matters more than most first-time managers realize.

**Kevin Baker**

> *The performance review is simultaneously a legal document, a development tool, an organizational signal, and a reflection of your leadership. Write it as if all four of those things are true at once, because they are.*

# What Every Review You Write Signals to Your Team

Your employees know more about how you conduct performance reviews than you might expect. They talk. They compare notes. Maybe not directly, but they do. And the collective impression your team forms about how you evaluate performance shapes the culture of your team in ways that are difficult to reverse once they are established.

| What the Review Does | What It Signals to the Employee | What It Signals to the Organization |
| --- | --- | --- |
| Reviews consistently specific and fair | My manager is watching What I do is seen and recorded accurately | This manager's assessments can be trusted in calibration and promotion decisions |
| Reviews consistently vague or generic | My manager is not paying attention the review reflects an impression not the work | This manager's ratings don't mean much they are inconsistent with what we observe |
| Reviews consistently inflated, everyone rated high | Performance doesn't actually matter here the rating is divorced from the work | This manager lacks evaluative courage Their top performers may be invisible to us |
| Reviews consistently harsh or deflated | My manager sees the negative and minimizes the positive. No one is good enough | This manager may be creating a retention problem and masking genuine talent. |
| Reviews delivered timely and followed up on | My manager takes the process seriously what was said in the review matters | This manager is engaged in development, not just documentation |

Look at the middle column, which shows what each pattern signals to the employee. Those signals accumulate over time. An employee who receives three vague reviews in a row draws a conclusion: the manager is not watching, or does not care, or cannot distinguish between good work and poor work. Any of those conclusions is damaging. The first produces an employee who stops performing for the review. The second produces disengagement. The third produces contempt, which is, from a management standpoint, the hardest thing to recover from.

## The Review and the Conversation — Why Both Matter and Neither Is Enough Alone

One of the most persistent misunderstandings about performance reviews is the relationship between the written document and the conversation that accompanies it. Most first-time managers treat the two as interchangeable: the review is the conversation, and the conversation is the review. They are not the same thing. They serve different purposes, and both are necessary.

| The Written Review: What It Does | The Review Conversation: What It Does |
| --- | --- |
| Creates the permanent official record | Creates the human experience of the feedback |
| Provides specific, documented evidence for ratings | Allows for dialogue, questions, and real-time response |
| Feeds compensation, promotion, and succession decisions | Builds or repairs the manager-employee relationship |
| Legally protects the organization and the employee | Enables the employee to process and internalize feedback |
| Can be referenced months or years later | Can be adjusted based on what the employee brings |
| Is read by HR, legal counsel, and senior leadership | Is experienced directly by the employee alone |
| Requires precision and permanence | Requires presence, listening, and flexibility |

The written review without the conversation is a document delivered into a void. The employee reads it, reacts to it alone, and either accepts or rejects it without any opportunity to engage with the person who wrote it. Whatever development was intended by the review is left entirely to the employee's interpretation. That interpretation may or may not match what the manager meant.

The conversation without the written review is a discussion that leaves no record. Whatever was said is subject to memory, the manager's and the employee's, and memory is an unreliable system. Feedback given only in conversation is the feedback most likely to be disputed, misremembered, or denied when it becomes consequential.

Both are required. The written review provides the anchor: the specific, documented, permanent record of what was observed and how it was assessed. The conversation provides the humanity, the chance for the employee to be seen and heard by the person whose assessment matters most to their career.

> *The form is the record. The conversation is the event. A great record delivered without a real conversation is paperwork. A great conversation without a great record is a memory. You need both. Every time. For every person on your team*

## What Happens When Reviews Are Done Poorly

The consequences of poor performance reviews are not abstract. They play out in specific, measurable ways: in retention data, in legal risk, in the quality of succession decisions, in team culture. Most managers who write poor reviews have no idea this is happening, because the consequences are diffuse and delayed. The employee who received three vague reviews and quietly disengaged does not leave a note explaining why. The lawsuit that arrives eighteen months after a termination does not announce itself in December when the review is filed.

Understanding what poor reviews actually cost is the most powerful argument for taking the time to do them well. Here is what the research, the litigation record, and the operational experience consistently show.

### The Cost to the Employee

- Receives inaccurate feedback and makes career decisions based on a distorted picture of their own performance
- Is denied the development opportunity that an honest, specific review provides
- Cannot defend themselves against future adverse actions because the record does not reflect what actually happened
- Loses trust in a manager who clearly was not watching, and often loses motivation alongside it
- May be under promoted because the review does not capture the full scope of their contribution
- In the worst cases: is terminated based on inadequately documented performance issues and has legitimate legal recourse

## The Cost to the Manager

- Loses credibility with HR and senior leadership who cannot trust their evaluations
- Increases legal risk through vague, inconsistent, or discriminatory language
- Fails to develop the talent on their team, which limits the team's capacity and the manager's own growth
- Misses the early warning signals that a well-written review forces them to name
- Finds themselves unable to support disciplinary actions or terminations because the documentation does not exist
- Builds a reputation as an evaluator whose assessments cannot be used, which limits their own advancement

## The Cost to the Organization

- Succession planning is built on unreliable data; the wrong people get promoted and the right ones are invisible
- Compensation decisions are disconnected from actual performance
- Legal risk accumulates with every inadequately documented termination
- High performers leave because they do not feel seen, and the reviews do not reflect what they actually contributed
- Low performers stay because the documentation does not exist to support the action needed
- Culture erodes as employees learn that the review is theater, and that performance does not actually determine outcomes

Read that last point again. When employees learn that performance does not determine outcomes, when the review is visibly disconnected from reality, the implicit contract of employment breaks. People stop working for the rating because the rating means nothing. They work for whatever actually gets rewarded in their specific environment, which may or may not be what the organization needs from them. That drift, invisible in any single review cycle, accumulates into a culture problem that takes years to correct.

## The Manager Who Gets This Right

This book is full of tools. Templates, phrase banks, language frameworks, legal guardrails, conversation scripts, documentation systems. All of it is useful. None of it is sufficient on its own. Because the manager who writes the review that actually matters, the one that gets kept, that shapes a career, that an employee references ten years later, is not the manager with the best template. It is the manager with the right understanding of what they are doing and why.

That manager understands that the review is not paperwork. It is a professional responsibility, as serious and consequential as any operational decision they make. They understand that every employee on their team deserves to be seen accurately, not flatteringly, not harshly, but accurately, because accurate assessment is the foundation of fair treatment and real development.

They understand that the review they write today will be read in contexts they cannot predict: in a promotion conversation, in a calibration session, in a legal proceeding, or simply on a hard night when an employee needs to remember that someone once saw what they were doing and put it on the record.

And they understand, perhaps most importantly, that getting this right is not only about the employee in front of them. It is about the kind of manager they are building themselves into, through each review, each conversation, each commitment to observe carefully and write honestly. That is the compounding investment this book is asking you to make.

## How This Book Is Organized: A Map Before You Begin

Before you proceed to Chapter 2, it is worth understanding how the pieces of this book connect, because they are designed as a system, not a collection of independent tips.

- Chapters 1 through 6 build the foundation. The purpose, the documentation habit, the self-review strategy, the rating system, and the language framework. Read these in order before you write your first review.
- Chapters 7 through 9 are the writing chapters. How to write for different performance levels, how to avoid legally dangerous language, and how to use the full review template. These are the chapters you will return to most often during review season.
- Chapters 10 and 11 cover the mid-year check-in and goal-setting, the two tools that make the annual review possible rather than improvised. (Note: goal-setting comes first in the actual performance cycle; Chapter 11 is placed after Chapter 10 for reference convenience.
- Chapters 12 through 14 cover the conversation. How to prepare, how to handle what you did not expect, and how to follow up in ways that make the review mean something after it is filed.
- The Conclusion and Appendices provide the closing philosophy and the reference tools: the Master Phrase Bank, the documentation calendar, the pre-submission checklist, and the HR/legal quick reference. These are resources you will return to throughout the year.

You do not need to read this book cover to cover before writing your first review. But it's worth reading Chapters 1 through 6 before opening the template. The tools in Chapters 7 through 9 are significantly more useful when you understand the foundation they are built on. Start here. Then build forward.

## Chapter 1 — Key Takeaways

• The performance review is simultaneously a legal document, a development tool, an organizational signal, and a reflection of the manager who wrote it. Write it with awareness of all four.

• Every review you write signals something to your team about what you observe, what you value, and whether performance actually determines outcomes in your organization.

• The written review and the review conversation are not the same thing. Both are necessary. The document provides the anchor. The conversation provides the humanity.

• Poor reviews have real, measurable consequences: for the employee who is not seen accurately, for the manager who loses credibility, and for the organization that makes decisions on unreliable data.

• The manager who gets this right is not the one with the best template. It is the one who understands what they are doing and takes it seriously enough to do it well.

• Read Chapters 1 through 6 before opening a template. The foundation matters.

# Chapter 2

# Building the File — Documenting Throughout the Year

*The performance review written in the final week almost always shows it. The one that reads with specificity, authority, and fairness was not assembled in December. It was assembled in pieces across twelve months by a manager who was paying attention.*

Every manager who has ever sat down to write a performance review one week before the deadline has felt the same thing: a sinking recognition that they are working from memory, and their memory is unreliable. They remember the big moments, such as the project that went sideways in October, the customer complaint in March, but the steady evidence of how an employee actually performed across fifty weeks of work? That record is gone. What replaces it is impression. And impression is a poor substitute for documentation.

This chapter is about building a documentation system that turns review writing into a retrieval exercise rather than a memory exercise. It is about the habit of capturing what you observe, briefly, consistently, and retrievably, so that when December comes, you are not guessing. You are selecting from a file of evidence that has been growing all year long.

The system does not need to be complicated. It needs to be consistent. Two minutes of documentation after a meaningful observation is worth more than two hours of reconstructed memory the week before reviews are due.

## Why Memory Fails — and Why That Matters

Human memory is not a filing cabinet. It is a narrative engine. It does not store events neutrally. Instead, it reconstructs them through the filter of what happened most recently, what felt most significant emotionally, and what confirms what we already believed about a person. For a manager writing performance reviews, this creates three specific and serious problems.

### The Recency Bias Problem

The event that happened two weeks before the review deadline will carry ten times more weight in an undocumented review than the event that happened in March. A strong employee who had a difficult November will receive a weaker review than their eleven months of excellent performance justifies. A struggling employee who pulled it together in December will receive a stronger review than their record supports. Both outcomes are unfair. Both are legally risky. Both are preventable with a habit of documentation.

## The Halo and Horn Effect Problem

A single powerful positive event, such as a project that saved the day or a customer who forwarded a compliment to leadership, can cast a halo over an employee's entire year in an undocumented review. A single significant negative event can cast a shadow. Documentation breaks both effects by giving you a body of evidence that is larger than any single moment. The pattern across twelve months tells the truth. One moment, no matter how dramatic, does not.

## The Consistency Problem

When managers rely on memory, they do not document their employees equally. They remember more about the employees they interact with most, the employees who cause the most friction, and the employees they like best. Meanwhile, the quiet, steady performer who never requires intervention and never causes problems will have the thinnest memory file at review time, even though their consistency may be the most valuable thing on the team. Documentation forces equity. If you are capturing observations for some employees and not others, you are introducing bias into the review process before you write a single word.

> *The legal dimension: In an employment dispute, the question is rarely 'what do you remember?' It is 'what do you have in writing?' A manager whose documentation is consistent, specific, and contemporaneous (recorded at the time of the event, not reconstructed later) is in a fundamentally stronger position than one who is relying on recall. Build the file all year. It protects your employees and it protects you.*

# What to Document — The Six Categories of Retrievable Evidence

Not everything that happens in a workday deserves to be documented. Part of building an effective documentation habit is learning to distinguish between routine events and the moments that will matter during review time. The following six categories cover the vast majority of what belongs in an employee's file.

### 01 Performance Against Goals and Metrics

- Goal milestone reached ahead of, on, or behind schedule, with the specific date and metric
- Productivity numbers that significantly exceeded or fell short of the established standard
- Quality metrics: error rates, accuracy percentages, customer satisfaction scores, any time a number moves meaningfully in either direction
- Attendance and reliability data: patterns of lateness, unscheduled absences, or conversely, perfect attendance through a high-demand period
- Budget or expense results: came in under, met, or exceeded the financial target and by how much

## 02  Project Involvement and Contribution

- Projects the employee was assigned to, led, or contributed to, with dates and their specific role
- Results of the project: what was the outcome, was it completed on time, what was the employee's measurable contribution
- Cross-functional projects that required the employee to work outside their normal scope
- Projects where the employee stepped up voluntarily beyond their assigned responsibilities
- Projects where the employee fell short of their expected contribution, named specifically

## 03  Observable Behaviors: Positive

- Moments when the employee demonstrated leadership, whether formal or informal, that you observed directly
- Times the employee received positive feedback from a customer, peer, or other manager. Capture the source and the substance
- Instances of going above and beyond: covering for a colleague, staying late to meet a deadline, solving a problem without being asked
- Behaviors that reflect company values: ownership, transparency, collaboration, safety consciousness
- Times the employee coached, mentored, or actively developed a colleague or direct report

## 04  Observable Behaviors — Developmental

- Specific instances where behavior did not meet the expected standard, with the date, what happened, and who was present
- Patterns you have noticed: the third time an employee interrupted a meeting, the second missed deadline, the recurring quality issue
- Coaching conversations you had: what was discussed, what the employee agreed to, what the follow-up was
- Moments where an employee's response to feedback was itself noteworthy: whether resistant, dismissive, or, conversely, exceptionally receptive
- Any verbal counseling or informal correction, with the date, subject, and what was said

## 05  Formal Events and Documentation

• Written warnings, counseling records, or disciplinary actions, with copies retained with the record

• Commendations, awards, or formal recognition: certificates, emails from leadership, and customer letters

• Training completed: course name, date completed, any relevant assessment results

• Certifications earned or renewed

• Formal coaching plans, PIPs, or improvement agreements, with every update and check-in documented

## 06  Significant Events and Context

• Operational events that affected the employee's performance: a system outage, a staffing shortage, an unexpected volume spike

• Personal circumstances disclosed by the employee that affected their work, noted carefully and without detail beyond what they chose to share

• Changes in scope, role, or reporting structure during the year

• Events that changed the goal landscape: a goal that became irrelevant, a goal that was added mid-year, a goal that was modified

• Moments of organizational change, including a new system rollout, a facility move, or a team restructuring, and how the employee navigated them

*A critical distinction: document behaviors and results, not interpretations. 'John arrived 12 minutes late on three occasions in March (3/4, 3/11, 3/19)' is documentation. 'John doesn't seem to care about being on time' is an interpretation. The documentation survives a legal challenge. The interpretation does not.*

**Kevin Baker**

# The Quick-Capture Format — Two Minutes That Change Everything

The reason most managers don't document throughout the year is not that they don't intend to. It is that they imagine documentation as a time-consuming, formal process. It is not. A useful performance note takes two minutes to write. It follows a simple format. And it is the difference between a review that is specific and defensible, and one that is vague and vulnerable.

The format is four elements. You do not need all four every time. But the more you have, the more useful the note will be six months later when you are reading it with fresh eyes trying to remember what actually happened.

| | |
|---|---|
| **DATE** | The exact date, not "sometime in the spring." March 14. October 2. The date is what makes a note retrievable and legally credible. |
| **EVENT** | What happened? One to three sentences. Describe the specific behavior or result. Who was involved? What was the context? |
| **IMPACT** | Why did it matter? What was the effect on the team, the customer, the operation, or the goal? This is what transforms a note into evidence. |
| **FOLLOW-UP** | What happened next? Was there a conversation? An agreement? A coaching moment? Did the employee respond? This closes the loop. |

Now look at the difference between a note that uses this format and one that doesn't. Both describe the same event. Only one is useful at review time.

| ✗ Too Vague: Avoid This | ✓ Retrievable: Write This Instead |
| --- | --- |
| *Maria had a safety issue in the pick zone. Talked to her about it.* | *March 7: Observed Maria operating a pallet jack in the pick zone without checking blind corner before proceeding. Near-miss with an associate exiting aisle 14. Spoke with her immediately following the incident. She acknowledged the lapse and agreed to the standard protocol. Documented in the safety log. Will monitor for recurrence.* |

| ✗ Too Vague: Avoid This | ✓ Retrievable: Write This Instead |
| --- | --- |
| *Carlos did a good job on the WMS rollout.* | *September 12: Carlos led the associate training component of the WMS rollout across two shifts. Trained 23 associates in four sessions over three days. Post-training error rate in receiving was 8% lower in the first week than projected. No associate escalations or complaints. Completed two days ahead of the implementation deadline.* |

| ✗ Too Vague: Avoid This | ✓ Retrievable: Write This Instead |
| --- | --- |
| *Derek seems disengaged lately.* | *November 3: Derek did not contribute during the weekly team meeting for the third consecutive week despite being directly asked twice for input on the new scheduling process. Peers have commented informally that the shift dynamic has changed. Scheduled a one-on-one for November 5 to address directly.* |

Notice what the right column has that the left column does not: a date, a specific behavior, the names of who was involved where relevant, the measurable impact, and the follow-up. Six months from now, the left column tells you almost nothing. The right column tells you exactly what happened, when it happened, and what you did about it.

# Where to Keep Your Notes — The System That Gets Used

The best documentation system is the one you will actually use consistently. A sophisticated system you abandon by February is worthless. A simple system you maintain all year is invaluable. Here are the options most managers find sustainable, from lowest friction to most structured.

## Option 1: The Email-to-Yourself Method (Lowest Friction)

After a meaningful observation, send yourself a quick email. Subject line: the employee's name and the date. Body: the four-element format above. At the end of each month, move those emails into a folder labeled by employee name. When review time comes, open the folder. Your notes are searchable, date-stamped, and already organized by person. This method requires no new tools, no new apps, and no behavior change beyond a two-minute email. Most managers who start with this method keep it indefinitely because it works.

## Option 2: The Shared Document Method

Keep a running Google Doc or Word document for each direct report, stored in a folder only you can access. Title each document with the employee's name and the review year. Add a new entry at the top (with newest entries first) every time you have something worth capturing. At the end of each quarter, spend fifteen minutes reading back through the entries and highlighting the ones that feel most significant. Those highlights become the backbone of the review.

## Option 3: The Physical Notebook Method

Some managers, particularly those who spend most of their time on the floor rather than at a desk, find that a physical notebook is the most reliable system. Keep a dedicated notebook for the year. Use a tab or section for each direct report. Write notes by hand immediately after an observation, before you move on to the next thing. The discipline of writing it by hand often produces more thoughtful entries than a quick digital note, and the physical notebook is harder to lose than a mental intention to document later.

## Option 4: Your Organization's HR System

Some HR platforms, including SuccessFactors, Workday, BambooHR, and others, include a continuous feedback or note-taking feature that allows managers to log observations directly in the employee's file throughout the year. If your organization has this capability and encourages its use, it is worth building into your habit. The advantage is that notes are already in the system when review time comes. The disadvantage is that entries may be visible to HR or the employee depending on how your organization has configured permissions. Know the rules before you use the feature.

## The Monthly 15: Your Documentation Maintenance Habit

Documentation is not just about capturing events when they happen. It is also about reviewing what you have captured, identifying gaps, and ensuring that your file is growing evenly across all of your employees, not just the ones who are most visible to you.

Once a month, set aside fifteen minutes. Put it on your calendar as a recurring meeting with yourself so it actually happens, then do the following:


- Open your notes for each direct report. Read through what you have captured since the last review.
- Ask yourself: Is this file growing? If you have nothing for this employee in the past four weeks, ask yourself why. Did nothing noteworthy happen, or were you not paying attention?
- Look for the quiet performer. The employee who never causes problems and never demands your attention is the one most likely to have a thin file. Make a note to observe them specifically in the coming month.
- Check the balance between positive and developmental notes. If a file has only positive observations, you may be overlooking growth areas. If it has only developmental notes, you may be overlooking genuine strengths.
- Flag any patterns you are beginning to see. The second late arrival. The third time a deadline was missed. The second time a peer mentioned a positive interaction with this employee. Patterns matter. Name them in your notes before you forget you noticed them.
- Add any context notes for the month, including operational changes, staffing challenges, volume spikes, or other factors that affected your team's performance. This context will matter when you are writing reviews and trying to explain why certain metrics looked the way they did.

Fifteen minutes. Once a month. Twelve times a year. That is three hours of maintenance documentation spread across the year. It is the difference between a review you are proud of and one you are embarrassed by.


## The Documentation Calendar — What to Capture and When

Some documentation is reactive: you capture what you observe when you observe it. While some documentation is proactive, where there are predictable moments in the performance cycle where you should be actively building your file. Here is a full-year calendar of those moments.

| Month | Priority Documentation Action |
| --- | --- |
| January | Set up individual files for each direct report. Document the goals agreed upon for this year: specific, measurable, with the success criteria written down. This is your baseline. |
| February | First monthly 15. Review your files. Note any early-year observations. Are goals being pursued? Is anything already off track? Document any early coaching conversations. |
| March | Q1 ends. Document Q1 metric results for each employee against their goals. Note any significant events from Q1. Flag any employees whose trajectory concerns you. |
| April | Review your file balance. Who has thin documentation? Intentionally observe those employees in April. Schedule brief informal check-ins if needed to gather more data. |
| May | Mid-year check-in preparation begins. Pull your notes from January through April. Identify the two to three most significant positive and developmental observations for each employee. This becomes your mid-year agenda. |
| June | Conduct mid-year check-ins. Document the conversation: what was said, what was agreed to, and any commitments made. Send the follow-up email. This entry is one of the most important in the file. |
| July | Q2 metric documentation. Note any significant second-quarter events. Flag any employees who are off track on mid-year agreements. Document any follow-up coaching conversations. |
| August | This is the month most managers fall behind. Maintain the Monthly 15. Do not let the summer create a documentation gap you will regret in December. |
| September | Q3 ends. Document Q3 results. Note trajectory changes: employees who have improved since the mid-year check-in and those who have not. The trajectory pattern across the full year is important context for your final assessment. |
| October | Begin light review preparation. Read through each employee's full file. Identify gaps. Note what is missing. Make a list of the things you still want to observe or confirm before the review cycle opens. |
| November | Active observation month. Fill the gaps in your files. Conduct any final coaching conversations that need to happen before year-end. Document those conversations contemporaneously. |
| December | Annual review writing begins. You are now working from a full file, not from memory. Your documentation work all year means the review is now a retrieval and selection exercise. Choose the most significant evidence. Write from what you have. |

**Kevin Baker**

> *The December manager who has been documenting all year does not dread review season. They open twelve months of organized observations, select the most significant evidence for each employee, and write with authority. The December manager who has been relying on memory opens a blank document and stares at it. Both had the same year. Only one was prepared to describe it.*

# Special Documentation Situations That Require Immediate Action

Most documentation can follow the routine described above. But certain events require you to stop what you are doing and write a thorough note immediately, within the hour, rather than waiting until that evening or the next morning. These are the situations where the documentation is not just useful for the review. It may be critical for the organization.

## Workplace Incidents and Safety Events

Any safety incident, whether a near-miss, injury, equipment misuse, or policy violation, must be documented immediately with full detail. Date, time, location, what happened, who was present, what was said, what action was taken. If a formal incident report is required by your organization, complete it. Then add a note to your own performance file for the employee involved. These two documents serve different purposes and both are needed.

## Verbal Warnings and Counseling Conversations

If you have a formal or informal correction conversation with an employee, document it the same day. Record what prompted the conversation, what you said, what the employee said, what was agreed to, and what the follow-up expectation is. This note may become the foundation of a formal PIP or a defense in a termination dispute. Write it as if someone else will read it—because they may.

## Exceptional Performance Events

When an employee does something genuinely exceptional, like solving an unexpected crisis, receiving unsolicited praise from a customer or senior leader, or delivering results that significantly exceed what was expected, document it immediately. The emotional specificity of 'I watched her walk a furious customer through a forty-minute order resolution and turn the conversation around entirely by the end' is only available the day it happened. Six months later, all you have is 'she handled a customer complaint well.' The first version is evidence. The second is an impression.

## When an Employee Discloses Something Personal

If an employee discloses a personal circumstance that may affect their work, such as a medical situation or a family crisis, that explains a change in behavior or performance, document the conversation carefully. Note what they disclosed (in general terms only. Do not record medical details), what accommodation, if any, was discussed, and what the expectation is going forward. Loop in HR immediately if there are accommodation implications. The performance file should note that a conversation occurred and what was agreed to. Personal details beyond what is operationally relevant do not belong in the file.

## Pattern Recognition: The Third Time Rule

When you notice something happening for the third time. For example, the third late arrival, the third missed deadline, the third time a peer mentions a conflict, document it explicitly as a pattern. Not just the third event, but the fact that it is a pattern. 'This is the third occasion in the past six weeks that this situation has occurred. Previous instances were documented on [dates].' That sentence, written in your notes, changes the nature of what you have from an isolated event to a behavioral pattern and patterns are what performance reviews and corrective action are built on.

# From File to Review — How to Use What You Have Collected

When review season arrives and you open your documentation file for each employee, you will have more material than you can use. That is exactly what you want. Your job at this point is not to document. It is to select, organize, and translate.

- Read the full file, start to finish, without writing anything. Just read. Let the year come back to you as a complete picture, not a collection of fragments.
- Identify the two to four moments or patterns that most define this employee's year. Not the most dramatic, but the most representative. What would a fair, objective observer say was the defining story of this person's performance?
- Group your evidence by review category: leadership, accountability, quality, communication, and so on. Which categories have strong evidence? Which are thin? Thin categories either need to be rated lower than you might prefer, or you need to acknowledge that you didn't observe enough to rate them confidently.
- For each category, select the one or two pieces of evidence that best support your rating. You do not need to include everything in your file. Just include enough to make the rating credible and the feedback actionable.
- Translate the evidence into review language. Your notes are raw material: specific, dated, and detailed. Your review language is the finished version, still specific and still grounded in evidence, but written in a tone and format appropriate for the official document.

| X Too Vague: Avoid This | ✓ Retrievable: Write This Instead |
| --- | --- |
| *NOTE: Oct 14: Pedro ran the quarterly safety walk with three associates. Found 6 violations in the pick zone. Had them corrected same day. Communicated results to the full team in the shift meeting.* | *REVIEW LANGUAGE: Safety leadership was evident throughout the year. During the October facility walk, six compliance issues were identified and corrected within the same shift, a response time that exceeded the standard. The results were communicated proactively to the full team, setting a clear example of the standard expected.* |

The note is the raw observation. The review language is the interpretation, grounded in the observation and elevated to the level of a professional assessment. You cannot write the second without the first. That is the entire argument for documentation throughout the year. The review does not create the evidence. It selects from it.

## A Note on Fairness

Documentation done well is one of the most equitable things a manager can do for their team. It forces you to observe everyone, not just the squeaky wheels. It replaces impression with evidence. It gives every employee, regardless of how loudly or quietly they perform, the same foundation of documented observation that their review deserves to be built on.

The employee who performs steadily and without drama, who never needs a corrective conversation and never creates a moment that demands your attention. That employee is the hardest to write a meaningful review for when you have been relying on memory. That employee deserves better than a generic review that sounds like it could have been written for anyone. Documentation is how you give it to them.

Start the file on day one. Keep it all year. Write from it in December. That is the whole system. Everything else in this book depends on it.

> ## Chapter 2 — Key Takeaways
>
> • Memory is not documentation. Impression is not evidence. The review written at the last minute almost always reads that way. The real work happens across the year.
>
> • Document behaviors and results, not interpretations. 'Arrived 12 minutes late on 3/4, 3/11, and 3/19' is documentation. 'Doesn't seem to care about punctuality' is an opinion.
>
> • The four-element format (Date, Event, Impact, Follow-Up) takes two minutes and is worth hours of recovered memory at review time.
>
> • Choose the system you will actually use: email-to-yourself, shared document, physical notebook, or HR platform. Consistency beats sophistication.
>
> • Document within 24 hours of the event. After 24 hours, memory has already begun to reconstruct. After a week, critical details are gone.
>
> • The Monthly 15, fifteen minutes of documentation review every month, ensures your files grow evenly across all employees and that patterns get named before they disappear.
>
> • Certain events require immediate documentation: safety incidents, counseling conversations, exceptional performance moments, and the third occurrence of any pattern.
>
> • When review season arrives, your job is to select and translate, not to remember. The file does the remembering. You do the writing.

# Chapter 3

# The Self-Review — The Most Underutilized Tool in the Performance Cycle

*Every employee knows something about their own work that their manager does not. The self-review is the formal invitation to say it. Most employees squander it with modesty, vagueness, or a copy-paste from last year. Most managers receive it and file it without reading it carefully. Both are leaving something significant on the table.*

Not every organization includes a self-review in its formal performance cycle. If yours does, use it. Fully, strategically, and with the same seriousness you bring to any professional document that will be read by the people who determine your trajectory. If your organization does not require one, this chapter will show you why it is worth doing anyway, and how to ask for the opportunity to submit one.

This chapter speaks to two different people, and it addresses both directly. First, the employee, specifically the first-time manager or rising professional who has been given a self-review form and does not know what to do with it. Second, the manager who receives self-reviews from their direct reports and has been treating them as a formality rather than a data source. The insight available in a well-read self-review is one of the most underappreciated tools in performance management. This chapter makes the case for taking it seriously, from both sides of the desk.

## Part One: For the Employee — Step onto the Podium

Here is the reality of how performance reviews are written. Your manager is responsible for multiple direct reports. They attend meetings, manage operations, handle escalations, and navigate their own performance expectations. They observe you, but they do not observe everything. They remember what was most visible, most recent, and most emotionally significant. The steady, important work you did in March that required no intervention and created no friction? There is a reasonable chance it is not in their notes.

The self-review is your opportunity to correct that. Not by complaining about what your manager missed. Not by inflating your contributions beyond recognition. But by stepping onto the podium, professionally, specifically, and with evidence, and making sure the record reflects what actually happened.

Most people do not do this. They write self-reviews that are either embarrassingly modest ('I tried my best and feel I contributed to the team') or uselessly generic ('I worked on several projects and met my goals'). Neither version serves them. Neither gives the manager anything they could not have written themselves without reading it.

> *The self-review is not bragging. It is professional advocacy. There is a meaningful difference between saying 'I am great' and saying 'Here is what I did, here is what it produced, and here is why it mattered.' The first is an opinion. The second is a record. Write the record.*

## What Your Manager Cannot Possibly Know Without You Telling Them

Before you write a word of your self-review, make this list. It does not need to be long. But it is the most important thing you will do to prepare.

- Work you completed that happened entirely within your area without escalating to your manager, because it went well and required no intervention
- Relationships you built or maintained with peers, customers, or cross-functional partners that contributed to organizational outcomes your manager may not have visibility into
- Problems you solved before they became problems. These are the issues you anticipated and addressed quietly, the customer complaint that never became a complaint because you caught it first
- Skills you developed this year, including courses, certifications, and other experiences, that did not require your manager's involvement and therefore may not be in their notes
- The full scope of a project where your manager saw the outcome but not the effort, the decision-making, or the obstacles you navigated along the way
- Contributions to your colleagues' success through coaching, covering, mentoring, and advocating, activities that are often invisible to the manager but deeply visible to the team

This list is not a grievance. It is the raw material of your self-review. Each item on it represents something worth capturing in the document, which is something your manager needs to know in order to write an accurate review and make fair decisions about your performance, your development, and your future.

## The Five Strategies for Writing a Self-Review That Gets Read

### 01  Lead with Impact, Not Activity

The weakest self-reviews describe what the employee did. The strongest describe what the employee produced. Activity without impact is a job description. Impact without activity is a claim. You need both, but lead with the result.

*Example: Not: 'I managed the inbound scheduling process.' Instead: 'I restructured the inbound scheduling window, cutting average appointment lead time from 72 hours to 38 hours, which directly improved our product availability metric by 14% in Q3.'*

### 02  Be Specific Enough to Be Undeniable

Vague self-reviews are easily dismissed. Specific ones are harder to argue with. Name the project, name the date, name the metric, name the people involved. Specificity is not arrogance; it is precision. And precision builds credibility.

*Example: Not: 'I played a key role in the system implementation.' Instead: 'I led associate training for the WMS rollout across two shifts, completing 23 associate certifications in four days, two days ahead of the go-live deadline.'*

### 03  Use the Language of the Organization's Values

Read your organization's review competencies before you write your self-review. Then write your accomplishments using the same language. If the organization values 'accountability,' show accountability. If it values 'customer focus,' connect your contributions to customer outcomes. This is not manipulation; it is alignment. The review attributes your manager must rate are the same attributes you should be reflecting in your self-assessment.

*Example: If the competency is 'Excellence: Anticipates future needs and challenges the status quo', write: 'I identified the gap in our IB receiving process three months before it created a service failure, proposed the fix, and drove its implementation.'*

### 04  Address Your Development Areas Before Your Manager Does

The most credible self-reviews include honest developmental reflection. When an employee names their own growth area, before the manager does, it signals self-awareness, which is itself a leadership competency. It also removes some of the sting from the manager's feedback, because the employee has already named the truth. Do not list every weakness. Name one or two real ones with a specific plan.

*Example: Not: 'I could improve my communication skills.' Instead: 'I want to develop my confidence in cross-functional settings where I am not the operational expert in the room. I am committing to leading one presentation at the Q3 manager summit to begin building that comfort.'*

## 05  Close With What You Want Next

The self-review is one of the few formal opportunities you have to tell your manager, on the record, where you want to go professionally. Use it. Not with demands or entitlement, but with a clear and specific statement of your aspirations. A manager who knows what you want is in a position to help you get there. A manager who does not know has no basis to advocate for you.

*Example: Not: 'I hope to continue growing with the organization.' Instead: 'I am ready to take on a cross-functional leadership responsibility in the next cycle. I would like to discuss what that could look like and how I can position myself for that opportunity.'*

# The Employee Self-Review Template

The following template is designed for the employee completing their own self-review. If your organization provides a specific form, use that form, but use the prompts below to prepare your thinking before you write a single word in the official document. The quality of your self-review is determined before you open the form, not while you are filling it out.

## Section 1 — Your Year in Your Own Words

### Opening Statement

▶ *In two to three sentences, describe the defining theme of your year. What kind of year was it, and why? Do not start with 'I worked hard.' Start with what the year actually produced.*

*Manager reads for: What the employee values most about their work. What framing they bring to their own performance.*

### Your Most Significant Accomplishment

▶ *Name the single most important thing you achieved this year. Not the thing you spent the most time on. Choose the thing that mattered most. Name what you did, how you did it, and what it produced.*

*Manager reads for: Where the employee places their own identity and pride. What they consider 'important' work vs. routine work.*

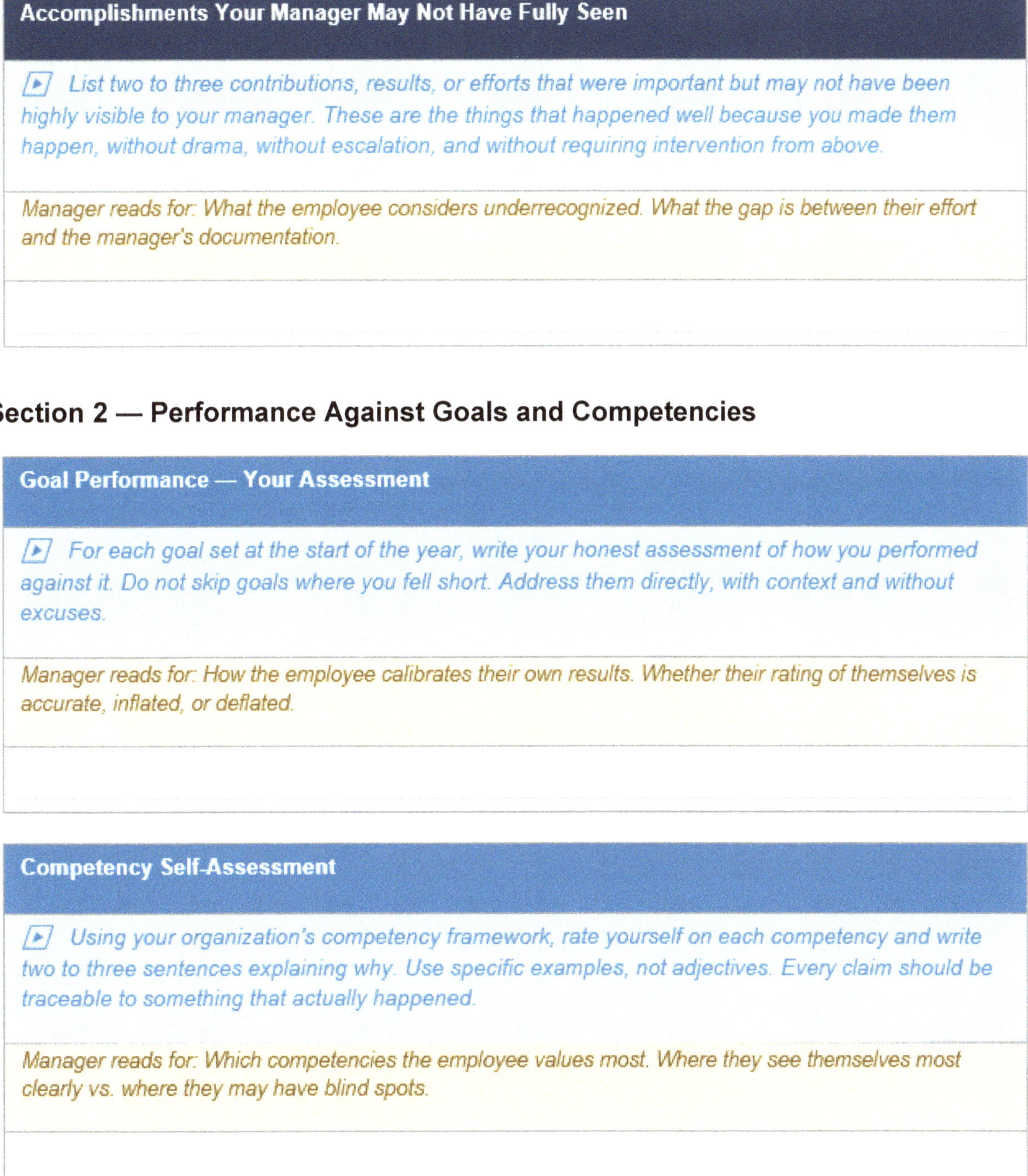

**Accomplishments Your Manager May Not Have Fully Seen**

List two to three contributions, results, or efforts that were important but may not have been highly visible to your manager. These are the things that happened well because you made them happen, without drama, without escalation, and without requiring intervention from above.

Manager reads for: What the employee considers underrecognized. What the gap is between their effort and the manager's documentation.

## Section 2 — Performance Against Goals and Competencies

**Goal Performance — Your Assessment**

For each goal set at the start of the year, write your honest assessment of how you performed against it. Do not skip goals where you fell short. Address them directly, with context and without excuses.

Manager reads for: How the employee calibrates their own results. Whether their rating of themselves is accurate, inflated, or deflated.

**Competency Self-Assessment**

Using your organization's competency framework, rate yourself on each competency and write two to three sentences explaining why. Use specific examples, not adjectives. Every claim should be traceable to something that actually happened.

Manager reads for: Which competencies the employee values most. Where they see themselves most clearly vs. where they may have blind spots.

## Section 3 — Growth and Development

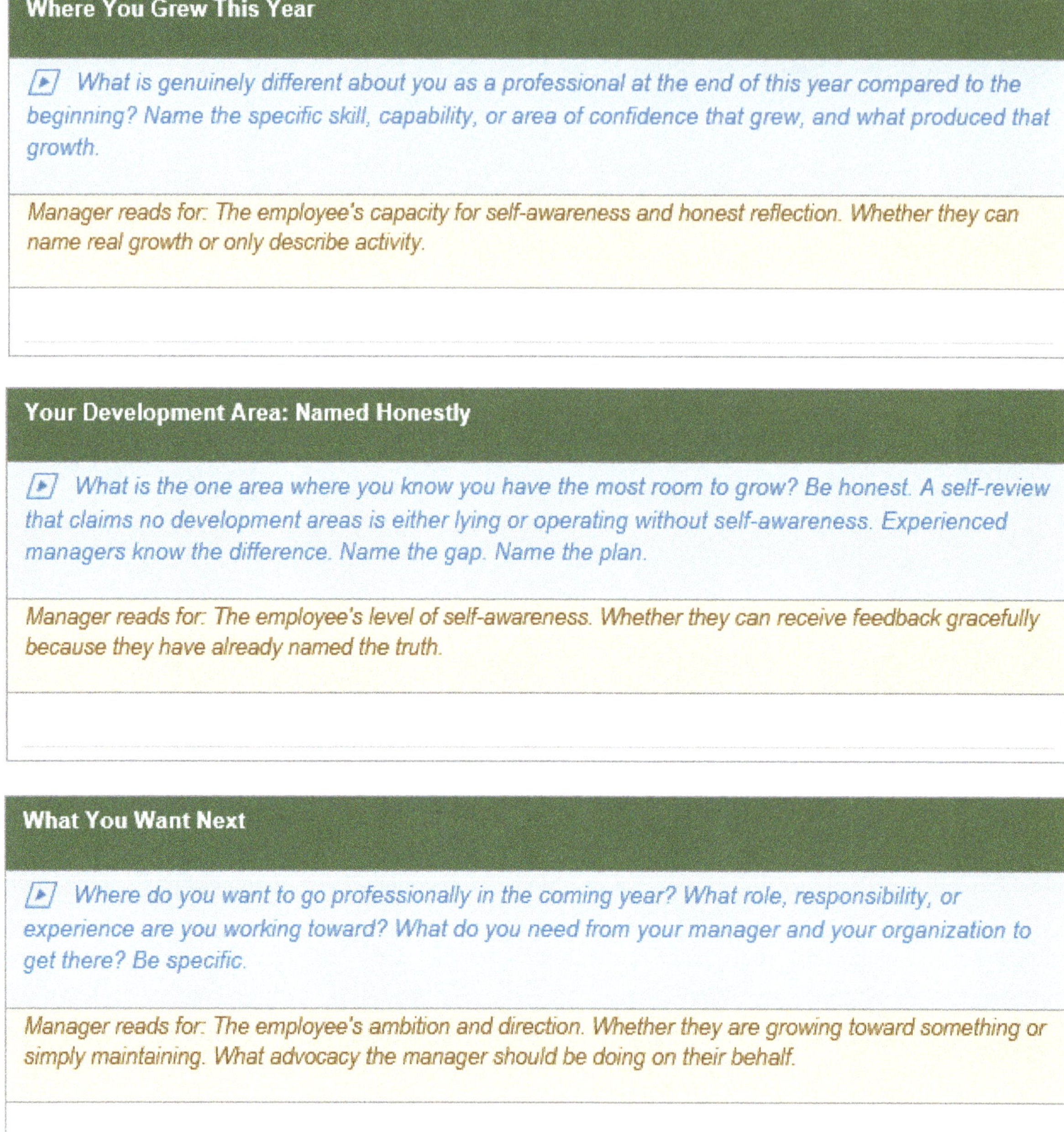

### Where You Grew This Year

*What is genuinely different about you as a professional at the end of this year compared to the beginning? Name the specific skill, capability, or area of confidence that grew, and what produced that growth.*

*Manager reads for: The employee's capacity for self-awareness and honest reflection. Whether they can name real growth or only describe activity.*

### Your Development Area: Named Honestly

*What is the one area where you know you have the most room to grow? Be honest. A self-review that claims no development areas is either lying or operating without self-awareness. Experienced managers know the difference. Name the gap. Name the plan.*

*Manager reads for: The employee's level of self-awareness. Whether they can receive feedback gracefully because they have already named the truth.*

### What You Want Next

*Where do you want to go professionally in the coming year? What role, responsibility, or experience are you working toward? What do you need from your manager and your organization to get there? Be specific.*

*Manager reads for: The employee's ambition and direction. Whether they are growing toward something or simply maintaining. What advocacy the manager should be doing on their behalf.*

**Kevin Baker**

## Section 4: Feedback for the Organization

**What Would Help You Perform Better**

*What resources, tools, processes, or support that, if they existed or improved, would make you more effective in your role? This is not a complaint section. It is an opportunity to contribute operational insight to the organization. Write it with that spirit.*

*Manager reads for: What barriers exist that the manager may be able to address. Where the organization is inadvertently making good performance harder.*

**Feedback for Your Manager**

*What could your manager do differently that would better support your performance and development? This section requires courage to write honestly and maturity to frame constructively. Do both.*

*Manager reads for: The state of the manager-employee relationship. Whether the employee feels safe giving upward feedback. What the manager may be doing that is limiting the employee's growth or engagement.*

# Part Two: For the Manager — How to Read a Self-Review

If you manage people who submit self-reviews, you are sitting on a data source that most managers never fully mine. The self-review is not just a list of the employee's accomplishments. It is a window into their psychology: what they value, what they believe about themselves, what they notice in their own work, and what they need from you. A manager who reads it carefully and responds to what is actually there will write a fundamentally better review and have a fundamentally better conversation than one who skims it and sets it aside.

## What You Are Really Reading For

When you open a self-review, you are reading two documents simultaneously. The surface document describes what the employee believes they accomplished. The underlying document reveals how they think about their work, their role, and their relationship with the organization. Both documents are valuable. The second one is more valuable.

- What does the employee consider their most important contribution? Is that consistent with what you consider their most important contribution? If not, that gap is the first conversation.
- What language do they use to describe themselves? Do they use the language of leadership, ownership, and impact, or the language of activity, effort, and intention? The gap between those two vocabularies tells you something about how the employee sees their role.
- What did they not mention? Every self-review has absences. The goal that was missed and not addressed. The behavioral gap that was not named. The relationship that went poorly. Silence is data.
- How do they frame their developmental areas? With specificity and ownership, or with vagueness and deflection? An employee who says 'I could improve my strategic thinking' has not named anything. An employee who says 'I struggled to see the downstream impact of my scheduling decisions and I am working to build that lens' has.
- What are they asking for, explicitly or implicitly? Career conversations embedded in self-reviews are often the clearest signal an employee sends about what they need and where they are headed.

> *The self-review gives you insight into what the employee feels is important. That insight allows you to get closer to their psychology and their values. And when you can mirror that, so that your review language reflects not just what they did but why it mattered to them. When that happens, the review stops being a document they receive and becomes a conversation they feel part of. That connection is what separates a review that is filed from a review that is kept.*

## The Psychology Mirror — Using the Employee's Language to Validate Your Assessment

This is the technique that transforms good reviews into exceptional ones. When you read the employee's self-review carefully, you will find words and phrases that reveal how they think about their work. The language of ownership. The language of service. The language of competition, or craftsmanship, or teaching. Whatever their dominant professional identity is, it will appear in how they describe themselves.

When your review language reflects that identity back to them, using their framework to validate their performance something important happens. The employee reads the review and feels recognized not just for what they did, but for who they are at work. That recognition is qualitatively different from a well-written assessment of results. It signals that you were paying attention to the person, not just the performance. And it creates a connection between the review and the employee's actual behaviors, values, and identity.

## Here is what that looks like in practice:

**Employee Self-Review: In Their Own Words**

*I take a lot of pride in making sure the team has everything they need before a shift starts. I don't like surprises and I don't like my people walking in unprepared. If I've done my job right, they can focus on the work instead of scrambling to figure out where we are.*

**Manager's Review: Language That Mirrors the Employee's Psychology**

*Your preparation discipline is one of the defining qualities of your leadership this year. The team in Seattle operates with a clarity and readiness at shift-start that is not accidental; it is the result of a supervisor who believes that preparation is a form of respect for the people you lead. That philosophy shows in every metric we track for this facility.*

▶ *What the Mirror Reveals: The employee values preparation as a form of care for their team. The review honors that framing: it does not just acknowledge the behavior, it validates the value system behind it. This employee will read that review and know their manager truly sees them.*

**Employee Self-Review: In Their Own Words**

*My goal this year was to make Burlington a place where people actually want to come to work. I know that sounds vague but it's real. When someone calls out, it costs us. When someone stays because they're proud of where they work, that's worth something. I tried to build that this year.*

**Manager's Review: Language That Mirrors the Employee's Psychology**

*The culture in Burlington this year was not an accident and it was not the result of any single initiative. It was the result of a leader who made a deliberate decision to make this building a place worth coming to. The associate opinion results, the retention data, and the energy visible on every facility visit are the measurable evidence of a value you have been living all year.*

▶ *What the Mirror Reveals: The employee values culture and belonging as a tangible operational outcome, not a soft concept. The review reflects that framing, treating culture as something measurable and real, which is exactly how this employee sees it. The mirror validates both the achievement and the values framework the employee used to pursue it.*

**Kevin Baker**

<table>
<tr><td>Employee Self-Review: In Their Own Words</td></tr>
</table>

*Honestly, I felt like I was playing catch-up most of the year in the strategic areas. The day-to-day I can do in my sleep. But when it comes to thinking three moves ahead and influencing things outside my building, I don't always trust my instincts.*

<table>
<tr><td>Manager's Review: Language That Mirrors the Employee's Psychology</td></tr>
</table>

*The operational excellence in the Texas facility this year is not in question; the metrics speak clearly. The development opportunity that remains is consistent with what you named yourself: the move from tactical mastery to strategic influence. You already know what you know. The next step is trusting that knowledge outside the four walls where you have already proven it.*

▶ *What the Mirror Reveals: The employee named their own developmental gap with unusual precision and honesty. The review honors that honesty by validating the self-assessment rather than softening it, and uses the employee's exact framing, 'trust your instincts outside your four walls,' to make the developmental challenge feel like a natural extension of existing strength rather than a criticism.*

In each of these examples, the manager's review language did not simply agree with the self-review or restate it. It elevated what the employee wrote, connected it to observable evidence, and reflected back the values framework the employee had revealed. The result is a review that the employee reads and thinks: my manager did not just observe me this year. They understood me.

## When the Self-Review and Your Assessment Diverge

Not every self-review will align with what you have observed and documented. Some employees will rate themselves significantly higher than you did. Others will undersell themselves. Both divergences are informative, and both require a response, though not the same response.

| What the Manager Observed | What the Employee Said in the Self-Review |
| --- | --- |
| *Manager assessment: Meets Expectations on leadership. Evidence: employee leads effectively inside the building but withdraws in peer group settings. Well documented across two rating periods.* | *Employee self-assessment: Exceeds Expectations on leadership. Evidence provided: strong team metrics, low turnover, positive associate survey results.* |

This divergence is not a contradiction to dismiss. The employee's evidence is accurate. Their team metrics are strong. Their associate survey results are real. The gap is that they are measuring a different dimension of leadership than you are. The review conversation becomes an opportunity to acknowledge both truths: what they are measuring and why it matters, and what you are measuring and why it also matters. The employee who understands that distinction, namely that there is a dimension of their performance they are not seeing clearly, will grow from that conversation. The employee who feels their self-assessment was simply overridden without discussion will not.

The rule is simple: when there is a significant gap between the self-assessment and your assessment, address it directly in the review meeting. Do not ignore it. Do not dismiss it. Do not apologize for your rating. Explain what you observed, acknowledge what they observed, and name clearly why your assessment landed where it did. That is a conversation. That is respect. And it is significantly more useful to the employee's development than a rating that arrives without explanation.

## Part Three: What to Do When Your Organization Does Not Require a Self-Review

If your organization does not include a self-review in the formal performance cycle, you have two options depending on which side of the desk you are on.

### For the Employee: Ask to Submit One Anyway

Most managers will welcome a well-written self-review even when it is not formally required, because it makes their job easier and their review more accurate. Send a brief email a week before review season asking whether you can submit a self-assessment for their consideration before they write your review. If the answer is yes, use this chapter's template. If the answer is no, bring your self-assessment to the review meeting as talking points for the employee self-assessment portion described in Chapter 12.

Either way, do the thinking. Prepare the material. Do not walk into a review meeting without having thought through your own year with the same seriousness your manager is bringing to the document they wrote about you. The employee who walks in prepared, knowing their accomplishments, knowing their gaps, knowing what they want next, and has a fundamentally different conversation than the one who walks in hoping it will go well.

### For the Manager: Ask for One Anyway

If your organization does not formally require self-reviews, ask your direct reports to complete one before review season. Provide the template from this chapter or a simplified version of it. Give them two weeks. Make clear that you will read it carefully and that their input will influence what you write.

The managers who build this practice into their teams, even when it is not required, consistently report two things. First, their reviews are more accurate because they have information they would not otherwise have had. Second, their review conversations are more productive because the employee arrives having already done serious reflection on their own year. Both outcomes are worth the fifteen minutes it takes to send the email asking.

**Chapter 3 — Key Takeaways**

• The self-review is the employee's formal opportunity to advocate for themselves. Most people waste it with modesty or vagueness. Use it with specificity and professional confidence.

• Lead with impact, not activity. Name what you produced, not just what you did. Specificity builds credibility.

• Address your development areas before your manager does. Self-awareness named first is strength. Self-awareness named defensively after feedback is a different conversation.

• Close the self-review with where you want to go next. A manager who knows your direction can advocate for you. A manager who does not know cannot.

• For the manager: the self-review is a psychological document, not just a checklist. Read it for what the employee values, what they notice in their own work, and what they need from you.

• The Psychology Mirror: when your review language reflects the employee's own values framework back to them, using their framing to validate your assessment, the review creates genuine connection between the document and the person.

• When the self-assessment diverges from your assessment, address it directly in the review conversation. Do not dismiss it. Explain your view, acknowledge theirs, and treat the gap as the conversation it deserves to be.

• If your organization does not require self-reviews, ask for one anyway. As a manager, request them from your team. As an employee, ask to submit one. The value is real for both sides.

Sometimes a review conversation does not go as planned. The employee gets emotional, pushes back hard, goes completely silent, or surprises you with information you were not prepared for. These moments are not failures of the review process. They are in the process of working. The manager who handles them with skill and composure earns trust that no well-written document can create on its own.

**Kevin Baker**

# Chapter 4

# Understanding the Rating System You've Been Given

*A rating without a shared understanding of what it means is not a rating. It is a number attached to a word attached to an assumption, and the assumption the manager is making may be entirely different from the assumption the employee is making when they read it. That gap, left unaddressed, is where most rating disputes begin.*

Every organization that conducts performance reviews uses a rating scale. Three-point. Five-point. Numerical. Descriptive. Competency-based. Whatever form it takes, the scale is the most visible and most misunderstood element of the entire review process.

Misunderstood in two directions. First, by managers who apply ratings inconsistently, using 'Meets Expectations' to mean very different things depending on the employee, the day, or how uncomfortable the honest rating felt. Second, by employees who interpret ratings without understanding what the scale was designed to mean, treating a 'Meets Expectations' as an insult when it was intended as a genuine acknowledgment of solid performance.

This chapter closes both gaps. It explains what rating scales are designed to do, how to apply them consistently and fairly, how to recognize and correct the grade inflation problem that distorts evaluations across most organizations, and how to calibrate your ratings against your peers so that a 'Meets Expectations' from you means the same thing as one from the manager down the hall.

## What Rating Scales Are Designed to Do

A rating scale is a standardization tool. Its purpose is to allow the organization to compare performance across different roles, departments, and managers using a common language. When a 'High Meets Expectations' from the Dallas facility and a 'High Meets Expectations' from the Seattle facility mean the same thing, the organization can use those ratings to make fair decisions about compensation, promotion, and succession.

When they do not mean the same thing, when one manager's 'Exceeds' is another manager's 'Meets', the scale fails at its purpose. Decisions made on inconsistent ratings are not fair decisions. They are decisions made on noise masquerading as signal.

This is why calibration exists, why HR audits rating distributions, and why the most experienced evaluators in an organization take the meaning of each rating level seriously. The scale only works if everyone using it applies it the same way.

> *A rating is not an expression of how you feel about an employee. It is a placement on a standardized scale that the organization uses to make decisions. Understand the scale before you use it, and apply it the same way every time, for every employee, regardless of how the rating makes you feel to give.*

## Decoding Your Scale — What Each Level Actually Means

The following table decodes the most common five-point scale used across organizations, including the LEAP framework used at American Hotel Register Company. (LEAP stands for Leadership, Excellence, Accountability, and Performance, a competency-based framework covered in full in Chapter 9.) If your organization uses a different scale, map the principles to your specific labels. The logic is the same regardless of the terminology.

| Rating Label | Plain Meaning | What Your Written Comments Must Show |
| --- | --- | --- |
| Exceptional / Exceeds Expectations | Far above standard | At least two to three specific examples of performance that went beyond what was required. Measurable impact named. Evidence that this employee performed at a level that was genuinely rare on your team or in the organization. |
| High Meets Expectations | Above standard, consistently | Evidence that the employee consistently performed above the baseline requirement across most areas. Not occasional excellence, but sustained performance above the norm. Metrics or behaviors that distinguish this from a standard Meets rating. |
| Meets Expectations | At standard, consistently | Evidence that the employee reliably met the requirements of the role across all essential areas. Note: this is a genuine positive rating, not a consolation prize. A team where everyone truly Meets Expectations is a high-functioning team. |
| Improvement Needed / Low Meets | Below standard in one or more areas | Specific named gaps: the behaviors or results that fell short of the expectation. Not a general impression of underperformance. Documented evidence. Clear statement of what the standard was and what occurred instead. |
| Unsatisfactory / Does Not Meet | Consistently below standard | Comprehensive documentation of sustained underperformance. Multiple specific examples across the rating period. Evidence that the gap was communicated and |

Read the third column carefully. What your written comments must show is not optional content; it is what makes the rating defensible. A rating without supporting narrative is an assertion. A rating with specific, evidence-based narrative is documentation. The organization can act on documentation. It cannot act on assertions.

## Meets Expectations Is Not an Insult — And Treating It Like One Costs You

This is the single most important correction in this chapter, and it is the one most first-time managers resist. In most organizational cultures, 'Meets Expectations' has acquired a connotation of mediocrity, of being average, overlooked, or merely tolerated. Employees brace for it. Managers feel guilty giving it. And so the inflation begins.

But here is what 'Meets Expectations' actually means when the scale is applied correctly: this employee reliably performed at the level the role requires across all essential areas. They showed up, did the work, met the standard, and can be counted on to do so again. That is not mediocrity. That is the baseline of professional performance, and in many organizations, it is rarer than it should be.

Consider what it means for an entire distribution network to have every manager rating 'Meets Expectations' as an accurate reflection of solid, reliable performance. That network is functioning. The work is getting done. The customers are being served. 'Meets Expectations' is a functional team.

### Why Grade Inflation Happens

Grade inflation, the pattern of rating employees higher than their performance justifies, is one of the most destructive forces in organizational performance management. It happens for predictable human reasons:

- It feels kind. Rating someone 'Meets' when you know they worked hard feels like a betrayal of their effort, even when the results did not match the standard.
- It avoids conflict. The conversation about a 'Meets' rating is easier than the conversation about 'Improvement Needed.' Inflation is avoidance wearing a compassionate mask.
- It protects the manager's relationship with the employee. A high rating produces gratitude. An accurate lower rating sometimes produces resentment.
- It reflects the manager's discomfort with their own authority. Giving a lower rating requires confidence in your assessment. Managers who doubt their own observation often inflate to avoid defending a position they are not sure of.

## What Grade Inflation Actually Costs

The costs of grade inflation are distributed and delayed, which is why managers rarely see them directly. But they are real and significant:

- The employee who is inflated never receives the accurate feedback they need to actually improve. They are praised into a false sense of their own performance level, and then blindsided when a termination or demotion arrives.
- The top performer on the team who genuinely Exceeds receives the same or similar rating as a peer who Meets at best. They notice. They draw conclusions. They leave.
- The organization cannot distinguish genuine high performers from adequate ones, which corrupts succession planning, promotion decisions, and development investments.
- The inflating manager loses credibility with HR and senior leadership who can see the pattern, and whose respect the manager needs when it matters.

Grade inflation is not kindness. It fails everyone: the employee who needed honest feedback, the team that deserved a fair standard, and the organization trying to make decisions it can trust.

The accurate rating is more useful to Maria than the inflated one. It tells her she is performing solidly and exactly where her next level of growth is. The inflated rating tells her everything is fine, which means she has no reason to grow.

| ✘ Grade Inflated — Harmful to Everyone | ✔ Accurately Rated — Honest and Useful |
|---|---|
| *Rating: Exceeds Expectations  Narrative: 'Maria has been a wonderful member of the team this year. She always has a positive attitude and goes above and beyond for her colleagues. She is a pleasure to work with and I am grateful to have her on the team.' Reality: Maria reliably met expectations in all areas. She did not exceed in any measurable way. The manager inflated because Maria is well-liked and the manager feared the conversation.* | *Rating: Meets Expectations  Narrative: 'Maria consistently met the performance expectations of her role this year. Productivity metrics averaged 94% of target across all four quarters. Her associate relationships are strong and she is a reliable presence during peak periods. The development opportunity for the coming year is taking more initiative in process improvement conversations. She has the operational knowledge; the next step is to share it.'* |

## Applying Ratings Consistently — The Three Questions

Before you assign a rating in any competency area, ask yourself three questions. If any of the three is hard to answer, that's a signal to pause before assigning the rating.

- What is the standard? What does 'Meets Expectations' look like for this specific competency in this specific role? Can you describe it in behavioral or metric terms that an objective third party would recognize?
- What is my evidence? What specific behaviors, results, or observations from this rating period support the rating I am about to assign? Not a general impression, but documented, specific evidence from the year.
- Would I apply this same rating to another employee at the same level with the same evidence? If you would rate Employee A differently from Employee B for the same observed behaviors, the discrepancy is not in the employees. It is in you.

The third question is the most important and the most uncomfortable. Inconsistency in ratings is where bias lives, not always intentional, not always conscious, but real and measurable. The manager who rates the employee they like more generously than the one they find difficult is introducing personal preference into a professional assessment. The manager who rates the employee from a different cultural background differently for the same communication behavior is introducing potential discrimination. The three-question test forces you to confront those inconsistencies before they become permanent record.

> *Consistency is not sameness. It does not mean every employee receives the same rating. It means the same standard is applied to every employee at the same level. Two employees can receive very different ratings for the same reason: their performance against the same standard was genuinely different. What cannot differ is the standard itself.*

## Calibration — Your Ratings Compared to Everyone Else's

Calibration is the process by which managers in the same organization review each other's ratings to ensure consistency across teams. It is one of the most valuable and most underutilized tools in performance management. If your organization conducts calibration sessions, participate in them fully. If it does not, understand what calibration is and calibrate informally with your peers.

## What Calibration Does

Calibration creates a shared understanding of what each rating level means in practice, not in theory, but in the actual cases being discussed. When a group of managers sit down and walk through their individual ratings for equivalent roles, the outliers become visible immediately. The manager whose ratings are consistently higher than their peers is either managing an exceptionally strong team (which the evidence should support) or inflating. The manager whose ratings are consistently lower is either holding a genuinely underperforming team or setting a standard their peers are not applying.

Calibration also surfaces the differences in what managers are observing and what they are counting as evidence. Two managers may give the same employee very different ratings for the same competency because they observed different behaviors, or because they weighed the same behaviors differently. The calibration conversation makes those differences visible and produces a more consistent standard across the organization.

## How to Prepare for a Calibration Session

- Know your ratings and be ready to defend each one with specific evidence, not general impressions. 'I rated her Exceeds because she is a strong performer' is not a calibration argument. 'I rated her Exceeds because she led the WMS rollout, trained 23 associates in four days ahead of schedule, and achieved a post-go-live error rate 8% below projection' is.
- Know the distribution of your ratings. If every employee on your team is rated Exceeds or High Meets, be prepared to explain why, and to have that explanation scrutinized. An unusually positive distribution requires unusually strong evidence.
- Enter the session with genuine openness to adjusting. The goal of calibration is not to defend every rating you gave. It is to arrive at a fair, consistent, organization-wide assessment. If the group's discussion reveals that your 'Meets' is equivalent to everyone else's 'High Meets,' adjust. That is the session working correctly.
- Pay attention to the ratings you receive feedback on. The competencies where your assessments diverge most from your peers are the competencies where you most need to examine your standards.

## Calibrating Before the Session — With Your Own Manager

Before submitting final ratings, share your draft assessments with your own manager. Not for approval, but for calibration. Ask specifically: are my ratings consistent with what you observe in my team? Are there any that seem significantly out of range with what you see from comparable managers? This conversation, held before the formal session, reduces surprises and gives you the opportunity to adjust with context rather than under pressure.

# When the Rating and Narrative Don't Match — The Most Common Mistake

The single most common rating error in performance reviews, and the one that creates the most legal and relational problems, is the mismatch between the written narrative and the assigned rating. This happens in two directions:

## Glowing Narrative, Mediocre Rating

The manager writes extensively about the employee's contributions, praises their attitude, describes their positive impact on the team, and then assigns a Meets or below rating. The employee reads the narrative and feels the rating is inconsistent with what was written. Their attorney reads the narrative and finds ample evidence to challenge the adverse action that followed.

If the narrative is genuinely positive and the rating is below what the narrative describes, one of two things is true: either the narrative is inflated and the rating is accurate, or the narrative is accurate and the rating is too low. Either way, they must be reconciled before submission. The narrative and the rating must tell the same story.

## Critical Narrative, High Rating

The manager writes about performance gaps, missed targets, and behavioral concerns, and then assigns a high rating because they are uncomfortable with the conflict, because they like the employee personally, or because they are hoping things will improve. This is grade inflation with a paper trail of contradictory evidence. The critical narrative makes the high rating indefensible and the high rating makes the critical narrative meaningless.

The rule is simple: write the narrative first, in accordance with the evidence. Then assign the rating that the narrative supports. Do not assign the rating first and then write a narrative around it. The narrative is the evidence. The rating is the conclusion. The evidence comes before the conclusion, every time.

## Chapter 4 — Key Takeaways

• Rating scales are standardization tools. Their purpose is to allow the organization to compare performance across roles and managers using a common language. They only work if everyone applies them the same way.

• Meets Expectations is not an insult. It means the employee reliably performed at the level the role requires. Applied accurately, it is a genuine positive rating that most organizations need more of.

• Grade inflation is not kindness. It denies the employee accurate feedback, corrupts organizational decision-making, and ultimately costs the inflating manager their credibility as an evaluator.

• Before assigning any rating, answer three questions: What is the standard? What is my evidence? Would I apply this same rating to another employee with the same evidence?

• Calibration, meaning comparing your ratings with your peers, is one of the most valuable tools in performance management. Participate fully. Enter with openness to adjusting. The goal is consistency, not defending every rating you gave.

• The narrative and the rating must tell the same story. Write the narrative first. Let the evidence determine the conclusion. Never assign the rating first and then write around it.

# Chapter 5

# The Language Framework Every New Manager Needs

*Vague language in a performance review is not a writing problem. It is a thinking problem. Writing 'great attitude' isn't a writing failure; it's a thinking gap. The question to ask is not 'what's a better phrase?' but 'what did I actually observe?' The manager who writes 'great attitude' failed to ask themselves what they actually observed, why it mattered, and what they want to see next. The phrase is the symptom. The framework is the cure.*

Most first-time managers write vague performance reviews for the same reason most people give vague directions: they know where they are going, but they have not thought carefully about how to describe the route to someone who does not already know the way.

The employee reading a review that says 'strong team player' or 'needs to improve communication' is receiving directions from a manager who has not thought carefully about the territory they are describing. The words sound like feedback. They contain no information the employee can act on.

This chapter gives you the framework that converts the territory you have been observing all year into language the employee can actually use. It is called the OID Formula, Observation, Impact, Direction, and it is the structural backbone of every piece of effective review language in this book.

| O  Observation<br><br>*What you saw or measured* | I  Impact<br><br>*Why it mattered* | D  Direction<br><br>*What comes next* |
|---|---|---|
| The specific behavior, result, or event that you directly observed or that is documented in your records. Names, dates, metrics, and context. | The effect of the observation on the team, the customer, the operation, or the organization. This is what transforms a fact into a meaningful assessment. | The forward-facing statement: what you expect to see continue, improve, or change. This is what makes the review a development tool rather than a grade. |
| **Ask yourself:**<br><br>*What did I observe? When? What was the context? Who was involved? What measurable result occurred?* | **Ask yourself:**<br><br>*What was the consequence of this behavior or result, for the team, the customer, the goal, or the operation?* | **Ask yourself:**<br><br>*What do I want this employee to keep doing, start doing, or do differently? What does success look like in the next cycle?* |

# The OID Formula — Observation, Impact, Direction

Every piece of effective performance review language can be broken down into three components. Not all three need to appear in every sentence, but every complete feedback statement needs all three somewhere in its vicinity. When any one of the three is missing, the feedback weakens.

Read the three questions in the bottom row carefully. They are the pre-writing prompts that most managers skip, and their absence is exactly what produces vague reviews. Before you write a single word about any competency, answer those three questions for the evidence you plan to use. The answers are your draft. The polished language comes after.

# The Formula in Action — Four Applied Examples

The following examples show the OID formula applied to four different performance situations: top performance, developmental feedback, safety, and team development. Each example shows how the three components combine to produce complete, specific, and useful feedback.

## Example 1 — Exceptional Performance (Leadership)

**Situation**

*DC Manager, Washington State: Leadership competency*

**O — Observation**

*In October, during my facility visit, three associates independently and without solicitation commented on how this building felt like a family, and on how proud they were to work for this manager and for the organization. The associate opinion survey from Q3 reflected the highest engagement scores in the network. Turnover in this facility over the past 18 months has been the lowest across all six locations.*

**I — Impact**

*A distribution facility where associates feel pride and connection does not happen accidentally. It is built, daily, by a leader who shows up the same way every day, who holds people accountable in a way that feels fair, who recognizes contribution in a way that feels genuine, and who creates an environment where people want to perform because they believe in where they work. The operational outcomes, namely retention, engagement, and productivity, flow directly from this cultural foundation.*

**D — Direction**

*This is the standard. Continue to be the leader this team has formed around. The next development opportunity is to take this capability outside the building, modeling the same deliberate culture-building with your peer group at the network level.*

**Rating This Supports**
Meets Expectations on Leadership, with specific developmental direction toward Exceeds

**Kevin Baker**

## Example 2 — Developmental Feedback (Quality Gap)

**Situation**

*DC Manager, Texas: Excellence / Quality competency*

**O — Observation**

*Inventory adjustment dollars for the Dallas facility in this rating period were $214,000 above the $600,000 annual target, finishing at $814,000. This represents a 35% variance against the agreed-upon goal. The top three error categories, bulk picking, put-away accuracy, and replenishment sequencing, were identified in Q2 and corrective actions were discussed in the July check-in. Error frequency in two of the three categories remained above baseline through Q4.*

**I — Impact**

*Inventory errors are not just internal metrics. Every adjusted dollar represents either a cost to the organization or a failure to deliver accurately to a customer. At the Dallas facility's volume level, a 35% variance above target has direct implications for operational margin and customer satisfaction. The corrective actions discussed in July were the right ones; the gap is in consistent execution at the supervisory level.*

**D — Direction**

*In the coming year, quality must be the primary operational focus for Dallas. The specific target is a 25% reduction from this year's $814,000 baseline, achieving $610,000 or below by year-end. Accountability at the supervisor level for error rates must be a standing agenda item in every leadership team meeting. I want to see a formal root cause analysis for each of the top three error categories by the end of Q1.*

**Rating This Supports**
Improvement Needed on Excellence / Quality

Kevin Baker

## Example 3 — Safety

**Situation**

*DC Supervisor, Illinois: Safety competency*

**O — Observation**

*On March 7, I directly observed an associate operating a pallet jack in the pick zone without checking the blind corner at aisle 14 before proceeding. A near-miss occurred with an associate exiting the aisle. This event was documented in the safety log. Two additional near-miss events were documented in this facility during the rating period, both in the pick zone.*

**I — Impact**

*Three near-miss events in a single facility during a single rating period is a pattern, not an anomaly. Near-misses are the leading indicator of recordable incidents. When they are addressed and corrected, the facility's safety performance improves. When they are logged and not meaningfully addressed, the next event is more likely to be a recordable injury. Safety is the one area where the consequences of inadequate performance cannot be undone after the fact.*

**D — Direction**

*Every near-miss in this facility must be followed within 24 hours by a documented root cause conversation with the associate involved and a communication to the full team. The safety walk process, which was inconsistently executed this year, must be conducted weekly with documented findings. Zero OSHA recordable incidents is the expectation for the coming year. This begins with treating near-misses as the serious events they are.*

**Rating This Supports**
Improvement Needed on Safety

## Example 4 — Positive Developmental Direction (Solid Contributor)

**Situation**

*DC Manager, New Jersey: Accountability competency*

**O — Observation**

*This manager owns every aspect of the Burlington facility: the successes, the gaps, and everything in between. When errors occur, the response is immediate and methodical: root cause is identified, process changes are implemented, and the team is communicated to clearly. Productivity in Burlington has improved year-over-year for the past three consecutive review cycles under this manager's leadership.*

**I — Impact**

*Accountability at this level, the kind that does not make excuses, does not diffuse responsibility, and does not wait to be told what to correct, is what builds a team that can be trusted with increasing scope. The consistent year-over-year improvement in productivity is the measurable proof of that accountability compounding over time. The team in Burlington performs at a high level because their leader holds them to one.*

**D — Direction**

*This is an area of genuine strength. The development opportunity connected to it is this: the same accountability this manager demonstrates inside the building needs to extend to the network level. Peer accountability, which means naming what is not working in cross-facility processes and driving solutions rather than adapting to the problem, is the next expression of this competency at the senior manager level.*

**Rating This Supports**
Exceeds Expectations on Accountability

Notice what each of these examples does not contain: adjectives without evidence, generalizations without specifics, praise without a corresponding observation, or criticism without a named standard. Every statement is traceable to something that happened: observed, measured, or documented. That traceability is what makes the OID formula the most important tool in this book.

## Why Vague Language Destroys Trust — and What It Really Says

Vague language in a performance review does not feel neutral to the employee reading it. It feels like evidence that the manager was not present for the year. And in most cases, that is exactly what it is.

When an employee reads 'great attitude' they do not think: how encouraging. They think: what does that mean? What did I do that produced this? How do I do more of it? They cannot answer those questions from 'great attitude.' They can only answer them from a specific observation that told them their attitude produced a specific outcome that mattered.

The table below shows the most common vague phrases and what they actually signal to an employee who reads them.

| The Phrase | What the Employee Hears | What OID Looks Like Instead |
|---|---|---|
| **Great attitude** | *My manager noticed I was pleasant but cannot name anything I actually did.* | *When the Q4 staffing shortage hit, you volunteered for three additional shifts without being asked. Your team noticed and said so. That is the attitude this building runs on.* |
| **Team player** | *I helped people sometimes. My manager has no idea what I specifically contributed to the group.* | *You covered two supervisors' responsibilities during the Burlington project without reducing output in your own area. That is not just cooperation; it is operational commitment.* |
| **Needs to improve communication** | *Something about how I talk to people bothered my manager. I have no idea what to change.* | *Written updates to your team on project status have been inconsistent. On four occasions this year, supervisors escalated confusion that a brief update would have prevented. The expectation is a weekly written status note during any active project.* |
| **Hard worker** | *My manager appreciates that I try. They cannot point to what I produced.* | *Your productivity numbers averaged 108% of target across all four quarters, the highest sustained rate in the network. That result does not happen without discipline and consistency that the rest of the team is watching.* |
| **Room for growth** | *My manager is hinting at something but will not say it directly. I am now anxious and unclear.* | *The development opportunity in this cycle is strategic thinking outside the day-to-day. You manage the current moment superbly. The next level requires managing the next quarter. That shift is the work of the coming year.* |

# Balancing Positive and Constructive — Why Both Matter and How to Use Each

Every performance review should contain both positive feedback and constructive feedback, for every employee, at every performance level. This is not about achieving an arbitrary balance for its own sake. It is about giving the employee a complete, accurate picture of where they stand.

A review that contains only positive feedback isn't an honest one. A review that contains only constructive feedback hasn't written a complete one. Both extremes fail the employee, and both create their own specific problems.

| Pattern | What the Employee Experiences | What the Manager Loses |
|---|---|---|
| All positive, no constructive | Feels good in the moment. Leaves without a clear sense of what to work on. Stops growing because the review implied there is nothing to grow toward. | Loses credibility as an evaluator. Top performers stop trusting the review because it is disconnected from reality. Grade inflation accumulates. |
| All constructive, no positive | Feels attacked, not assessed. Becomes defensive or disengaged. Loses motivation because the review contained no evidence that the work was seen. | Damages the relationship. Loses the employee's willingness to hear the constructive feedback, which is buried under resentment. |
| Balanced — honest positive and honest constructive | Feels genuinely seen, for both what worked and what did not. Has something to build on and something to work toward. Trusts the review as an accurate reflection of the year. | Builds credibility as a fair evaluator. Creates the foundation for the employee to receive harder feedback in the future because the good work was also named. |

The practical guideline: for every developmental area you name, name at least one genuine strength. Not to soften the developmental feedback, but to complete the picture. An employee who knows what they do well and what they need to work on has everything they need to grow. An employee who only knows what they need to work on knows half the truth.

**The One Thing Most Managers Do Wrong With Positive Feedback**

They rush past it. They spend two sentences on the accomplishments, then pivot to 'however, there are areas for improvement' and spend the next four paragraphs there. The employee stops reading the positive section because they have learned that it is just the setup for what comes next.

Positive feedback deserves the same specificity and space as constructive feedback. If an employee did something exceptional, describe it with the full OID formula, observation, impact, direction, before you move on. Let it land. The employee who feels genuinely recognized for what they did well is far more open to hearing about what they need to do differently. That is not strategy. It is human psychology.

## The Pre-Write Test — Before You Type a Single Word

Before writing any feedback statement, try answering these five questions. If any are difficult, that's a signal to go back to the documentation file first.

- What specifically did I observe? Can I name the date, the behavior, the result, or the metric?

- Why did this matter? What was the effect on the team, the customer, the goal, or the operation?

- What do I want this employee to do next? Keep doing this? Improve on it? Apply it differently?

- If I remove all the adjectives from what I am about to write, is there still content? If not, I have written an impression, not a feedback statement.

- Would I stand behind this statement in a room with HR, senior leadership, and the employee present? If not, I need to revise it. Not soften it, revise it.

The last question is the most important. The review you are willing to read aloud in any professional setting is the review you should submit. The one you would be uncomfortable defending is the one that still needs work.

# Vague vs. Specific — The Difference in Practice

| ✗ Vague — Useless to Everyone | ✓ Specific — OID Applied |
|---|---|
| *Shows great initiative.* | *Without being asked, identified a gap in the IB receiving process that was causing a 4-hour delay in product availability. Proposed a modified dock assignment sequence, piloted it over two weeks, and reduced the delay to under 90 minutes. Initiative in this case meant seeing the problem, building the solution, and proving it worked.* |

| ✗ Vague — Useless to Everyone | ✓ Specific — OID Applied |
|---|---|
| *Struggles with time management.* | *On six occasions during this rating period, project deliverables were submitted after the agreed deadline. In four of those cases, the delay was not communicated in advance. The expectation going forward is either delivery on time or advance communication when a deadline is at risk, not both missed after the fact.* |

| ✗ Vague — Useless to Everyone | ✓ Specific — OID Applied |
|---|---|
| *A valuable member of the team.* | *During the Burlington facility move, this manager contributed 11 days of on-site support from the Dallas operation, at direct cost to their own team's metrics, because the organization needed it. The Burlington move was completed on time in part because of that commitment. That is what organizational value looks like.* |

## Chapter 5 — Key Takeaways

• The OID Formula, Observation, Impact, Direction, is the structural backbone of all effective review language. Every complete feedback statement needs all three components.

• Before writing any feedback, answer three questions: What did I observe? Why did it matter? What do I want next? The answers are your draft.

• Vague language is not a writing problem; it is a thinking problem. The phrases that feel like feedback ('great attitude,' 'needs to improve communication') contain no information the employee can act on.

• Every review needs both positive and constructive feedback. Do not rush past the positive. Let specific recognition land before pivoting to developmental areas.

• The pre-write test: if you remove all the adjectives from your draft, is there still content? If not, you have written an impression, not feedback.

• The review you would be willing to read aloud in any professional setting is the review you should submit. The one you would not defend is the one that still needs work.

# Chapter 6

# Word Banks and Phrase Libraries by Category

*A phrase bank is not a shortcut. It is a starting point. Every phrase in this chapter becomes useful only when you attach it to specific evidence from your documentation file. A phrase without evidence is still a vague review. A phrase with evidence behind it is the beginning of a great one.*

A phrase bank is not a shortcut but a starting point. These phrases help managers in selecting clear, professional language for performance reviews, but they are effective only when backed by concrete evidence from your documentation. Lacking evidence, a phrase may appear generic; with evidence, it becomes specific, credible, and useful.

This chapter provides review language organized by performance category, performance level, and workplace context.

Use these phrases as scaffolding, not scripts. Read through each category before review season. Highlight the phrases that resonate with what you observed in your employees this year. Then open your documentation file, find the specific evidence that supports the phrase, and write the two together. The phrase gives you the direction. The evidence makes it real.

Each category is organized into two columns: language for strong or recognized performance, and language for developmental or improvement-needed performance. Both columns are necessary. A phrase bank that only covers positive language produces inflated reviews. One that only covers developmental language produces incomplete ones.

## 01  Performance and Productivity

Use this language when assessing how effectively the employee produced results against operational targets, including cartons per hour, error rates, same-day shipping, budget adherence, or any quantifiable metric specific to their role.

| Strong Performance Language | Developmental Performance Language |
|---|---|
| • *Productivity metrics averaged [X]% of target across all four quarters, the highest sustained rate in the network this year.*<br><br>• *Consistently met or exceeded daily output expectations, demonstrating operational discipline that is visible in the data.*<br><br>• *Achieved the [specific metric] target in [timeframe], ahead of the pace set in the prior cycle.*<br><br>• *Performance in this area has improved year-over-year for [X] consecutive cycles, a trajectory driven by this leader's consistent standards.*<br><br>• *Delivered results in [specific area] that placed this facility at the top of the network ranking for the full year.*<br><br>• *When volume spikes occurred, output was maintained without a corresponding increase in errors. That is a measure of operational control that is genuinely rare.*<br><br>• *Budget adherence throughout the year was [X]%, demonstrating financial discipline alongside operational results.* | • *Performance metrics in [specific area] finished the year at [X]% below target. The gap between current output and expectation must close in the coming cycle.*<br><br>• *Productivity was inconsistent throughout the year: strong in [period] but below standard in [period]. Sustained performance at the expected level is the work of the coming year.*<br><br>• *The standard for this role is [specific metric]. For the last [X] months, this standard has not been met with regularity.*<br><br>• *Errors and quality misses in [specific area] occurred at a rate above what is acceptable for a facility of this scope.*<br><br>• *Cost management in [specific area] finished above budget by [X]%. Understanding and managing expenses within the established parameters is a requirement of the role.*<br><br>• *Output was adequate in low-demand periods but has not consistently met expectations during peak. Sustained performance across all volume conditions is the expectation.* |

## 02  Quality and Accuracy

Use this language when assessing inventory accuracy, order accuracy, error rates, rework, miss-ships, mis-picks, and any metric related to the correctness of work produced.

| Strong Quality Language | Developmental Quality Language |
| --- | --- |
| • Inventory accuracy in this facility has improved by [X]% from the prior year baseline, the result of systematic process improvements and consistent accountability. | • Inventory adjustment dollars finished the year at $[X], which is [X]% above the stated target of $[X]. Quality improvement in this area is the primary operational focus for the coming year. |
| • Error rates in [specific area] are among the lowest in the network and reflect a team that understands accuracy is a service commitment, not a metric. | • Error rates in [specific area] remain above acceptable levels. The corrective actions discussed in [period] have not yet produced the improvement expected. |
| • The same-day shipped percentage finished the year at [X]%, at or above the network expectation for every quarter. | • Mis-ships and mis-picks in this facility occurred at a frequency that directly impacted customer satisfaction. This must be addressed at the process and accountability levels. |
| • Quality in the inbound processing area improved measurably after the supervisor training in Q2. That investment is visible in the results. | • Quality of work produced is not yet at the level of consistency this role requires. The gap between current output and the expected standard is [specific description]. |
| • Mis-ship rates have trended down for [X] consecutive quarters, demonstrating that the process changes implemented in [period] are holding. | • Accuracy issues have persisted in [specific area] despite conversations in [period] about the expectation. A formal root cause analysis and corrective plan is required by end of Q1. |

## 03  Communication

Use this language when assessing how effectively the employee communicates upward, downward, and laterally: in writing, in person, in group settings, and across functions.

| Strong Communication Language | Developmental Communication Language |
|---|---|
| • *Articulates complex operational thinking in both writing and conversation with a clarity that builds confidence in the people receiving it.*<br><br>• *Communicates expectations to the team with a directness that removes ambiguity and enables consistent execution.*<br><br>• *Written updates are organized, timely, and contain the context the recipient needs to act on, without requiring follow-up questions.*<br><br>• *In peer settings, contributes perspective that is substantive, well-reasoned, and delivered with enough confidence that it shapes the direction of the conversation.*<br><br>• *Listens with the same quality of attention that is given to speaking, which is rare and which makes this person a trusted communicator at every level of the organization.*<br><br>• *When delivering difficult information, whether to the team or to senior leadership, does so directly and without deflection.* | • *Written communication has not consistently met the standard expected at this level. Updates are often incomplete, delayed, or require the recipient to follow up for context.*<br><br>• *In group settings with peers or senior leaders, the contribution level drops significantly from what is demonstrated inside the building. The development opportunity is to trust that perspective in any room.*<br><br>• *Expectations communicated to the team have not always been specific enough to enable consistent execution. The coming year requires more precision in how direction is delivered.*<br><br>• *Communication when things go wrong, including service failures, process gaps, and team issues, has been slower and less direct than the role requires. Proactive communication of problems is a non-negotiable expectation.*<br><br>• *There is a gap between what is known and what is shared. Information that would benefit peers or the leadership team is frequently held rather than communicated. This limits the collective.* |

## 04  Teamwork and Collaboration

Use this language when assessing how the employee contributes to the collective, including their team, their peer group, and cross-functional partners, beyond their individual responsibilities.

| Strong Teamwork Language | Developmental Teamwork Language |
|---|---|
| • *Willingly shared expertise and resources with the peer group, including [specific example], at direct cost to their own team's metrics. That is organizational commitment.*<br><br>• *Built working relationships with [specific peers or partners] that have produced measurable operational results, including the PPI, the WMS rollout, and the carrier compliance program.*<br><br>• *When a peer needed support, the response was immediate and genuine. Collaboration at this level is noticed and valued.*<br><br>• *Accepts responsibility for group outcomes, not just individual ones. When the network misses, does not look for the cause to be elsewhere.*<br><br>• *Peers seek this person's input because the input is useful, not because of the title. That credibility is built. It is not assigned.* | • *The contribution to the peer group outside the four walls of [facility] has not matched the strength of the operational contribution inside them. The next level of leadership requires both.*<br><br>• *Cross-functional relationships with [sales/HR/transportation/other] have been strained this year. Words and plans have not yet translated into the results those partners need. Action is what builds trust in those relationships.*<br><br>• *In group settings, participation has been passive when the role requires active contribution. Remaining in the back of the pack in peer settings is not a neutral choice. It is a missed opportunity to lead.*<br><br>• *Collaboration requires more than availability. It requires actively sharing what you know before you are asked. That proactive contribution is the development area for the coming cycle.* |

## 05  Reliability and Attendance

Use this language when assessing schedule adherence, attendance consistency, follow-through on commitments, and the degree to which the employee can be counted on, particularly for hourly and front-line supervisory roles.

| Strong Reliability Language | Developmental Reliability Language |
|---|---|
| • *Attendance and schedule adherence have been without exception this year. The team operates with confidence that this supervisor will be present, on time, and prepared at the start of every shift.* | • *Attendance and schedule reliability did not meet the standard expected of this role during this review period. On [X] occasions, the shift began without adequate supervision due to late arrival or unplanned absence.* |
| • *When a commitment is made, whether to a deadline, a standard, or a follow-up, it is kept. That consistency builds trust faster than any single performance event.* | • *Commitments made in coaching conversations and one-on-ones have not always been followed through. The expectation is that when something is agreed to in this organization, it happens.* |
| • *During the [peak period/project/staffing shortage], was present and reliable when other supervisors were not. That reliability is what peak season requires and what good managers remember.* | • *The team has absorbed the impact of inconsistent availability from this supervisor. Associates cannot perform at their highest level when they are unsure whether supervision will be present.* |
| • *Follow-through on assigned responsibilities has been complete and timely throughout the year. Delegating to this person means the work gets done.* | • *Reliability is not just attendance. It is the confidence of everyone around you that what you said will happen will actually happen. That confidence must be rebuilt in the coming cycle.* |

## 06  Leadership Potential and Growth Trajectory

Use this language when assessing an employee's readiness for the next level of responsibility, whether they are a high-potential associate being considered for supervision, a supervisor being assessed for management, or a manager being evaluated for senior leadership.

| Strong Leadership Potential Language | Developmental Leadership Potential Language |
| --- | --- |
| • *The informal leadership this person demonstrates with peers and associates is the clearest predictor of their readiness for the next formal level.*<br><br>• *Has voluntarily taken on responsibilities outside the defined scope of the current role, not when asked, but because the team needed it and this person saw it.*<br><br>• *Associates respond to this person the same way they respond to the formal supervisor. That is leadership before the title, and it is exactly what the title requires.*<br><br>• *The questions this person asks in one-on-ones and development conversations are the questions of someone who is already thinking at the next level.*<br><br>• *Ready-in-12-months is the assessment for succession in this role. The evidence: [specific observation or result that supports readiness].* | • *The technical competency is present. The leadership disposition, meaning the willingness to own outcomes, set direction, and hold others accountable, is still developing.*<br><br>• *Leadership readiness requires more than skill. It requires the confidence to make decisions when the right answer is not obvious and to stand behind them. That confidence is the development area.*<br><br>• *Peer influence, the ability to shape how others think and work without formal authority, is the gap between this person\u2019s current level and the next. It is a skill that can be developed with intentionality.*<br><br>• *The succession timeline for this role is currently assessed at ready-in-24-months. What needs to change to move that to 12: [specific development activities].* |

## 07  Safety and Compliance

Use this language when assessing safety performance, OSHA recordable rates, near-miss reporting, safety walk compliance, and adherence to company policies and regulatory requirements. Safety language requires particular specificity. Vague safety assessments are both operationally and legally inadequate.

| Strong Safety Language | Developmental Safety Language |
| --- | --- |
| • *Zero OSHA recordable incidents for the full year. This result reflects a culture where safety is a value, modeled by the leader and held by the team.*<br><br>• *Weekly safety walks were conducted and documented without exception. Corrective actions were taken same-day and communicated to the full team.*<br><br>• *Near-misses were reported, documented, and addressed with root cause conversations within 24 hours, exactly the process that prevents near-misses from becoming recordables.*<br><br>• *Associates in this facility operate with a safety consciousness that is visible from the first moment of any visit. That culture is not trained in. It is led in.*<br><br>• *Achieved Chasing Zero [medal level] recognition, a network standard that this facility reached through consistent behavioral discipline, not luck.* | • *Safety performance this year, with [X] OSHA recordable incidents, is above the acceptable level and must improve. Zero recordables is the expectation, not the aspiration.*<br><br>• *Near-miss events in this period were documented but not consistently followed up with root cause conversations or team communication. That gap is where the next recordable event lives.*<br><br>• *Safety walk compliance was inconsistent. The process exists to catch hazards before they cause harm. When the walk does not happen, the hazard is not caught.*<br><br>• *Associates in this facility require a stronger safety message from their supervisor. Safety behavior is modeled from the top. The current model is insufficient.*<br><br>• *A formal safety improvement plan is required for the coming cycle, with weekly accountability check-ins and documented corrective actions for every violation found.* |

## 08  Growth, Development, and Continuous Improvement

Use this language when assessing how actively the employee is investing in their own professional development, adapting to change, and looking for ways to improve the work they do.

| Strong Growth Language | Developmental Growth Language |
| --- | --- |
| • *Completed [course/certification/development activity] and applied the learning directly to [specific operational change]. Development was not an event. It was an investment with a visible return.* | • *Professional development activities agreed to at the start of the year have not been completed. The IDP is a commitment, not a suggestion.* |
| • *Is a continuous learner in the truest sense: asking questions, challenging assumptions, and bringing new thinking to problems that have been solved the same way for years.* | • *Adapting to [new system/process/organizational change] has been slower than the role requires. The expectation is that change is met with engagement, not resistance.* |
| • *When given feedback, the response is action, not acceptance, not defense, but measurable adjustment in the behavior or result that was named.* | • *Feedback that has been given consistently this year has not yet produced the behavioral change expected. The gap is not in the feedback. It is in the follow-through.* |
| • *Has grown more in this rating cycle than in any prior cycle. The trajectory is the story, and the trajectory is excellent.* | • *The trajectory this year was flat when the expectation was upward. The coming year must show evidence of growth in [specific area], not effort toward growth, but evidence of it.* |
| • *Self-awareness at this level, the ability to name one\u2019s own growth areas honestly and specifically, is itself a leadership competency.* | • *This role requires intellectual curiosity and a hunger for improvement that has not been consistently visible this year. Finding that drive is the most important work of the coming cycle.* |

## 09  Operations and Distribution — Context-Specific Language

The following phrases are drawn specifically from distribution, logistics, and operations management contexts, covering the language of the warehouse, the DC, and the supply chain. They are designed for the manager whose team works in an operational environment and whose reviews need to reflect that reality specifically.

### Inbound and Receiving Operations

- *IB appointment scheduling was managed to the [X]-hour window established at the start of the year, keeping the dock clear and product moving.*
- *Reduced product arrival-to-availability time from [X] hours to [X] hours through a modified dock assignment sequence, a change that improved downstream pick efficiency.*
- *Inbound processing accuracy improved by [X]% following the associate training in Q[X]. The investment in training is visible in the result.*
- *The inbound team consistently absorbed volume spikes without service impact, a sign of preparation, flexibility, and strong supervisory direction.*
- *IB errors, including product damaged, miscounted, or misrouted at receiving, occurred at [X rate] during this period. Reducing this to [X target] is the operational priority for Q1.*

### Outbound and Shipping Operations

- *Same-day shipped percentage for the year was [X]%, [above/at/below] the [X]% network expectation. [Above: This result reflects the discipline built into the daily operating rhythm. Below: Closing this gap is the primary service focus for the coming year.]*
- *The 3:00 PM order cut-off was met with the zeal and operational focus it deserves, as associates understood the commitment and operated accordingly.*
- *Shipping accuracy in this facility finished the year at [X]%, with [specific context and trend].*
- *Mis-ships in [specific area or period] required escalation and customer communication. The root cause, [specific cause], must be addressed systematically, not case by case.*
- *Outbound productivity (cartons per hour processed) averaged [X] for the year, [above/at/below] the [X] target.*

## Inventory and Warehouse Management

- *Inventory shrink and adjustment dollars were managed within [X]% of budget, a reflection of tight process discipline in receiving, put-away, and cycle counting.*
- *The cycle count program was executed on schedule throughout the year, producing accurate variance data that allowed for proactive correction rather than reactive discovery.*
- *Slotting improvements implemented in [Q/period] reduced low-pick frequency by [X]% and contributed to a [X]% improvement in pick efficiency.*
- *Inventory accuracy at year-end physical was [X]%, [above/at/below] the [X]% target established at the start of the year.*
- *Bulk organization and end-cap replenishment in this facility created a measurable improvement in pick path efficiency, driven by this supervisor's operational instincts.*

## Technology and Systems (WMS, Voice, TMS)

- *Demonstrated a rapid and accurate understanding of the WMS implementation. The learning curve was shorter than projected, and the contribution to the go-live timeline was significant.*
- *Was an early adopter and active advocate for the voice-picking technology, which accelerated adoption on the floor and reduced the ramp-up time for the full team.*
- *Used system reporting capabilities to identify operational patterns and drive decisions, a data-driven approach that elevated the quality of day-to-day management.*
- *System adoption on this team was [X]% by [date], ahead of the implementation schedule and a direct result of effective supervisor-level training and accountability.*
- *Technology proficiency is still developing. Full utilization of [WMS/voice/TMS] capabilities would significantly improve the quality of operational decision-making in this facility.*

## Chapter 6 — Key Takeaways

- A phrase bank is a starting point, not a shortcut. Every phrase becomes useful only when attached to specific evidence from your documentation file.
- Both columns of every category are necessary: strong performance language and developmental language. A bank that only covers positive phrases produces inflated reviews.
- Before using any phrase, confirm you have the specific observation that supports it. If you don't, find a phrase that matches the evidence you have, or return to your documentation file.
- Operations-specific language, meaning the vocabulary of receiving, shipping, inventory, and systems, is what makes a distribution manager's reviews read as authoritative rather than generic.
- The metric is the evidence. A review that names a specific number, such as same-day at 98.5%, adjustment dollars at $614,000, or CPHW at 42, is a review that is specific, defensible, and credible.
- Return to this chapter at the start of every review season. Read through each category. Highlight what resonates. Then open your documentation file and find the evidence that brings the phrase to life.

# Chapter 7

# Writing for Different Performance Levels

*The biggest mistake managers make is writing the same review for everyone. Your top performer deserves more than a generic 'great job.' Your struggling employee deserves more than vague criticism. Every person on your team is owed specific, honest, useful language.*

After years of reviewing managers at every level, from first-year supervisors to seasoned distribution center leaders, the pattern is always the same. Managers who struggle with reviews don't struggle because they don't know their people. They struggle because they don't know how to translate what they know into written language that is specific, fair, and actionable.

This chapter gives you the language tools to write for four distinct performance situations: your top performers, your solid and steady contributors, your employees who are improving but not yet there, and your employees who are genuinely struggling. Each section includes real-world examples drawn from reviews written over two decades of leading distribution operations, anonymized to protect privacy but unchanged in substance.

Read all four sections even if you think you only need one. Understanding how language shifts across performance levels will make you a significantly better writer across the board.

## Writing for Your Top Performers

Here is a truth most managers don't realize until it's too late: your top performers are the most under-reviewed people on your team. When review season comes, managers spend the majority of their mental energy on difficult conversations, including the struggling employee, the improvement plan, and the uncomfortable rating. By the time they get to their best performer, they're drained, and the review shows it.

Your top performers notice. High performers are, by definition, people who pay close attention and hold themselves to high standards. When their review reads like a form letter, with vague praise, no specifics, and a rubber-stamp "exceeds expectations," they feel it. It tells them their manager isn't watching. And managers who aren't watching eventually lose the people they can least afford to lose.

Strong reviews for top performers do three things: they name what the person specifically did, they explain the impact of that work on the team or organization, and they open a door to what comes next. Even an exceptional employee has a next level. Your job is to point toward it while making them feel genuinely seen.

## The Language of Top Performance

Notice what the following example does not say: 'great job,' 'hard worker,' 'positive attitude,' or 'team player.' These phrases appear in thousands of reviews every year and mean almost nothing. What follows is what specific, evidence-based top-performer language actually looks like.

---

*REAL-WORLD EXAMPLE — DC MANAGER, WASHINGTON STATE | PERFORMANCE COMPETENCY | EXCEEDS EXPECTATIONS*

*In every area of performance, this facility is operating at an above-average or greater level.*

*Productivity, service, housekeeping, and safety are all exemplary. You are masterful at seeing the moves on the chessboard before others do. You proactively remove obstacles and know when to take risks. You communicate extremely well — not only with peers but also with associates across the network — to gain an understanding of issues and to manage accordingly.*

*The associates in this facility follow you willingly because they respect you, and because they trust you as both a person and as a leader. Several associates, without solicitation, commented on how it felt like a family and how proud they were to work for this organization. That culture does not happen by accident. It is built deliberately by a leader who shows up the same way every day.*

*Great job this year. The work has been exceptional.*

---

Let's break down why this works. First, the manager names specific performance categories, including productivity, service, housekeeping, and safety, rather than just saying "performance was excellent." Second, it uses a concrete metaphor: "seeing the moves on the chessboard." That phrase does more work than "strategic thinking" ever could. Third, it attributes the building's culture directly to the manager's daily behavior. It tells the employee not just what they achieved but how they achieved it, and why that matters.

Finally, notice the closing. It's short, warm, and direct. Top performers don't need long, elaborate compliments. They need you to say clearly: I see what you're doing. It's working. Keep going.

*Key principle: Specificity is respect. When you name what someone actually did, you prove you were paying attention. That proof matters more than any rating score.*

## Balancing Praise with a Developmental Thread

Even your highest-performing employees should leave a review with something to work toward. The goal is not to manufacture criticism; it's to treat your top performers as the professionals they are. High performers want to grow. A review with no developmental component feels like a ceiling, not a compliment.

The key is to frame the development as an extension of existing strength, not as a gap or a correction. Watch how the following example introduces a growth area for the same top-performing manager without diminishing the Exceeds rating.



This is a Meets Expectations on a Leadership competency for someone who received an Exceeds on Performance. The manager is not being inconsistent; they're being precise. The employee is exceptional at one thing and still developing at another. The review honors both truths. It does not soften the developmental feedback. It names the exact behavior to change, specifically stepping up without being nudged, and it ends with an encouraging reframe: your existing skills are transferable. Use them.

## Phrases That Work for Top Performers

Use these as starting points. Customize each with specific evidence from your own observations.

- You proactively identify and remove obstacles before they become operational problems.
- The culture in this building reflects your leadership style in every meaningful way.
- Your team follows you because they believe in you, not because they have to.
- You communicate with a level of clarity and specificity that raises the standard for those around you.
- The results this year did not happen in spite of the challenges; they happened because of how you led through them.
- You have built something here that will outlast any single performance cycle.
- Your willingness to take ownership of failures as readily as you accept credit for wins sets you apart.
- Others seek your input not because you have the title, but because you have earned the trust.

# Writing for Your Solid, Steady Contributors

The middle of your performance distribution is where most of your team lives, and it is the hardest group to write for. Top performers give you clear material. Struggling employees give you specific problems to document. But the solid, steady contributor, the person who shows up, does the job well, doesn't cause problems, and doesn't set the world on fire, presents a different writing challenge entirely.

The danger with this group is two-fold. First, the temptation to write something generic because everything is, technically, fine. Second, the temptation to manufacture improvement areas that don't really exist just to make the review feel substantive. Both mistakes communicate the same thing to the employee: I didn't think very hard about you.

A strong review for a solid contributor is honest about what "meets expectations" means, which is more than most managers realize. Meeting expectations consistently, across all responsibilities, over an entire year, is a real achievement. It deserves to be named as one.

## Naming What 'Meets Expectations' Actually Means

> **REAL-WORLD EXAMPLE — DC MANAGER, TEXAS | COMMUNICATION COMPETENCY | EXCEEDS EXPECTATIONS**
>
> *You have done a masterful job engaging with your peer group and have built real confidence with them. You have done this by articulately expressing your thoughts and ideas both in writing and orally.*
>
> *You have displayed strength in your abilities and shared your wisdom openly. Your peer group — as well as I — have been impressed by your willingness to listen, learn, and act.*
>
> *I expect this trust with your peers to build throughout the coming year. Continue to share, and they will actively seek you out for guidance.*

Note what this review does: it names the behavior ('articulately expressing thoughts and ideas both in writing and orally'), names the audience impact ('your peer group has been impressed'), and ends with a forward-looking statement that functions as both recognition and expectation. The employee knows exactly what they did, why it mattered, and what to keep doing.

> **REAL-WORLD EXAMPLE — DC MANAGER, TEXAS | PROBLEM SOLVING | MEETS EXPECTATIONS**
>
> *You use a data-driven approach as your primary method to break down processes and find underlying root causes. This can be difficult when the person collecting data does not truly understand it. You have demonstrated that you can do this well. You are methodical and logical.*
>
> *I would like to see you probe beyond the first right answer you arrive at. Often there is a more right answer. Do not be too quick to find data that supports your hypothesis — at times this can lead to a bias in your analysis.*
>
> *You have a very desirable skill set. Use it to its full potential. At times I can see you settling. Do not settle.*

This is Meets Expectations with teeth. It acknowledges the strengths, specifically that the employee is methodical, logical, and data-driven, and then pushes on the limitation with precision. "Probing beyond the first right answer" is specific. "Do not settle" is direct. Together they give the employee something concrete to work on without making the review feel like a criticism of their character.

> *A Meets Expectations rating is not a consolation prize. Written well, it tells an employee: you are reliable, you are valued, and here is exactly what your next level looks like.*

**Kevin Baker**

**Phrases That Work for Solid Contributors**

- Your consistency is one of your most valuable professional qualities. The team counts on you.
- You have demonstrated that you can be trusted with significant responsibility and have delivered on it.
- You approach your work with integrity and a seriousness of purpose that is evident to everyone around you.
- Your willingness to learn and absorb new information has been a consistent strength this year.
- Where this rating period showed room for growth, the opportunity is not about working harder; it is about working with more strategic intention.
- You have built the foundation. The next step is to build on it.
- Your team reflects your standards. Continue to raise them.

# Writing for Employees Who Are Improving but Not Yet There

This is the employee who is trending in the right direction but hasn't arrived yet. Maybe they're new to the role. Maybe they've taken on a bigger scope. Maybe they've turned a corner after a difficult period. Whatever the context, the writing challenge is the same: how do you give an honest assessment that acknowledges real progress without overstating it, and without deflating someone who is genuinely trying?

The answer is sequencing. Start with what you have witnessed improve. Name it specifically. Then name what still needs work, with equal specificity. Then end with a clear statement of confidence or direction. The employee should finish reading and know: I'm making progress. Here's exactly where I still need to grow. My manager believes I can get there.

**REAL-WORLD EXAMPLE — DC MANAGER, TEXAS | LEADERSHIP COMPETENCY | MEETS EXPECTATIONS (YEAR TWO)**

*You are able to influence others and gain their commitment rather than their mere compliance. This is done by being a democratic leader who asks for and relies on others' input. The team is strengthened when the leader shows that each person is valued and respected as an expert in their position.*

*It has been a challenging year. There was a period during this rating cycle where you struggled to find your leadership footing. However, in the second half of the year you hit your stride, and the metrics bear this out.*

*As a leader you must continue to challenge yourself and your team. There are times when you must be an autocratic leader — be aware of when that is needed and react accordingly.*

*Do not let your own self-created barriers get in the way of your growth or your team's optimal performance levels.*

This review is honest about a difficult period without dwelling in it. The pivot to the second half of the year and the metric improvement gives the struggle context and shows it as temporary rather than defining. The closing line about self-created barriers is the kind of specific, personal feedback that only comes from a manager who is paying close attention.

## Multi-Year Narrative: Tracking Growth Across Review Cycles

One of the most powerful things you can do as a reviewing manager is show an employee their own growth over time. When your reviews build on each other, when this year's feedback references last year's development areas, it tells the employee that you are tracking their journey, not just grading a moment in time.

The following two examples are from the same employee, reviewed two years apart. Notice how the developmental theme from the first review evolves in the second.

**REAL-WORLD EXAMPLE — DC MANAGER, ILLINOIS | 2017 REVIEW | LEADERSHIP | MEETS EXPECTATIONS**

*You have demonstrated your ability to influence your team through your words and actions. You are an inspirational leader that your team gravitates toward and responds to. It is not uncommon for your leads, admin, and others to feel empowered to speak up and share their opposing views. This is only achieved when a leader gives the right amount of autonomy and genuinely values the input of others.*

*Outside of your building, your comfort level and confidence take a step back. Do not doubt your importance at all levels of this organization. Your ability to lead from the front does not diminish when you walk out the front door.*

**REAL-WORLD EXAMPLE — DC MANAGER, ILLINOIS | 2020 REVIEW | LEADERSHIP | EXCEPTIONAL/EXCEEDS**

*You stepped back into leadership right in the middle of a pandemic. The organization as a whole was anxious and in disarray. Everything was upside down. You returned and brought with you a soothing voice and a stabilizing presence.*

*The facility became the rock of the organization. It remained open and operating after everything else had closed. It became the bright light at the end of a dark tunnel for many during turbulent times.*

*You did this through effective leadership: great communication to your team, empathetic listening and decision-making, and by modeling stability and calmness under pressure. The building represents the life of this organization, and you have protected it and kept it safe with your leadership.*

Three years separate these two reviews. In 2017, the feedback was: you lead brilliantly inside your building, but you shrink when you walk out the door. By 2020, that same manager had become the stabilizing force for the entire organization during its most difficult moment. The growth arc is visible because the reviews were written with enough specificity to make the comparison meaningful.

This is why your notes matter. This is why you track behavior throughout the year rather than just the week before reviews are due. The review that names real growth is the most powerful development tool you have.

> *When an employee sees their own growth reflected back to them in your language, something shifts. They stop seeing the review as an evaluation and start seeing it as evidence. Evidence that someone was watching. Evidence that they are becoming something. That is the review that gets kept.*

### Phrases That Work for Improving Employees

- The second half of this year demonstrated what you are capable of when you trust your instincts.

- You have taken actionable feedback and converted it into measurable improvement. That is not easy to do.

- The trajectory matters as much as where you are right now. The direction is right.

- You have shown the ability to course-correct under pressure, which is one of the most important leadership skills there is.

- The foundation you built this year is what next year will be built on. Do not underestimate it.

- Growth is not always linear. What matters is that you kept moving forward when it would have been easier not to.

## Writing for Employees Who Are Struggling

This is the hardest section to write, both in a review and in a book. Not because the language is complicated, but because most managers avoid it. They soften, they hedge, they bury the real message under so many qualifiers that the employee walks out of the meeting genuinely unsure whether there's a problem.

That is a failure of leadership, and it has consequences. When an employee who is not meeting expectations is told, in writing, in their official review, that things are going reasonably well, you have taken away their ability to choose. They can't fix a problem they don't know they have. And when it escalates to discipline or termination, they will pull out that review and point to it as evidence that you never told them.

Writing honest, direct feedback for a struggling employee is not unkind. It is the most professional thing you can do. Done right, it is also the most human thing. It says: I respect you enough to tell you the truth. I believe you can do something with it. Here is exactly what needs to change.

## The Structure of Honest Developmental Feedback

When writing for a struggling employee, follow this structure:

- Name the expectation clearly, including what was required, not what the employee fell short of
- Name the specific behavior or result that missed the mark, using evidence, not adjectives
- Name the impact, explaining why this matters to the team, the customer, or the organization
- Name the path forward, describing what specifically needs to change, and by when if possible
- Close with a statement of investment, confirming that you are still willing to support the work

Notice what is not in that list: character judgments, emotional language, comparisons to other employees, or vague directives like 'needs to improve attitude.' Every word in a struggling employee's review should be something you could read aloud in an HR meeting and stand behind completely.

REAL-WORLD EXAMPLE — DC MANAGER, TEXAS | WORK PRODUCTIVITY & QUALITY | MEETS EXPECTATIONS (EARLY TENURE)

*Overall, I believe this is the area with the most opportunity for growth during this rating period. This competency is closely associated with end results — metrics. In some areas there is significant room to reach new heights.*

*Quality and overall productivity are not where they need to be for a facility of this size and scope. The budget target for next year has been established, and there is every expectation it will be met. To accomplish this, the facility must average 52 cartons per hour processed for the entire year. For the last several months, this standard has not been met with any regularity.*

*As the leader, you must set the standard and drive expectations in performance. Make the job easier for your team by demanding process improvements, holding poor performers accountable, and refusing to fall into the excuse trap.*

*Own the building, own your leadership team, and own the results.*

This is a struggling area within an overall Meets Expectations review for a newer manager. What makes this effective: the standard is named specifically (52 cartons per hour), the gap is named without blame (it has not been met with regularity), the action is named clearly (demand improvements, hold poor performers accountable, avoid excuses), and the closing is a charge, not a threat. 'Own the building' is motivating. 'Or else' is not.

## When the Review Precedes a Formal Performance Improvement Plan

If an employee is approaching the threshold for a formal PIP, the language in their annual review must be unambiguous. Do not soften it because you feel bad. Do not soften it because you like the person. Do not soften it because you hope things will turn around before anyone notices.

The review must say clearly: performance has not met expectations in the following areas. It must name those areas specifically. It must name the consequences of continued gaps where appropriate. And it must be written in language that HR, legal counsel, and a third party could read without confusion.

If you are writing a review that may precede formal disciplinary action, make sure you review your organization's HR guidelines and consult with your HR partner before writing a single word. Legally risky language and phrases that create liability deserve careful attention before you put anything in writing. For more details, see Chapter 8.

## Phrases That Work for Struggling Employees

- Performance in this area has not consistently met the expectations established at the start of the review period.
- The gap between current results and required results must be closed in the coming cycle. This is not optional.
- The standard is clear. The path to meeting it is also clear. What is needed now is consistent execution.
- Effort alone is not the measure. Results are the measure. The results in this area are not yet where they need to be.
- This feedback is given because the expectation for this role is high, and you have the capability to meet it.
- The work ahead is significant. I am invested in your success and will provide the support needed, but the accountability belongs to you.

# Before You Start Writing: A Pre-Review Checklist

Before you sit down to write any review, regardless of performance level, run through this quick checklist. It will save you from the most common mistakes.

- Do I have at least three specific examples of behavior or results to draw from for this employee?

- Am I writing about what the employee did, or what I felt about what they did?

- If I remove all adjectives from this draft, is there still substance left?

- Does my rating match my written comments? (A glowing review with a Meets rating, or a critical review with an Exceeds, can create confusion and legal risk)

- Have I named at least one specific development area for every employee, including top performers?

- Would I be comfortable reading every sentence of this review aloud in a meeting with HR present?

- Have I given this employee something concrete to do with this feedback?

If you can answer yes to all seven, you are ready to write. If any of them give you pause, go back to your notes before you open the document.

> *The best review you will ever write is not the one with the most words. It is the one the employee reads and thinks: my manager actually knows me. That is what you are aiming for, every time, for every person on your team.*

# Chapter 8

# The Phrases That Will Get You in Trouble

*A note before we begin: This chapter reflects general best practices in performance review language and employment communication. It is not legal advice. Employment law varies by state, by industry, and by the size and structure of your organization. What applies in California may differ from what applies in Texas, and what applies to a 500-person company may differ from what applies to a 50-person one. Your organization's HR team and legal counsel are the authoritative source on compliance in your specific context. When something in this chapter raises a question for your situation, bring it to them before you submit. That's not a hedge; it's how this is supposed to work.*

*A performance review is a legal document. It may not feel like one when you're sitting at your desk trying to find words for someone who drives you crazy, but it is. The day it gets pulled into a grievance, a termination hearing, or a lawsuit, every word you wrote will be read by people you have never met, looking for things you never intended to say.*

Most managers who write legally problematic reviews don't do it intentionally. They do it because they're tired, because they're trying to be kind, because they're venting frustration they haven't processed, or because nobody ever taught them where the lines are. This chapter is about the lines.

This chapter covers something else too: the role of Human Resources in the review process. Because this is where HR earns its stripes. Not in the policy memos, not in the onboarding paperwork, but here, in the audit of the language a manager chose when they thought nobody important was watching. A good HR professional is the last line of defense between a well-meaning manager and a very expensive mistake. Understanding what they're looking for makes you a better writer and a better partner.

## Why Your Review Is Already a Legal Document

The moment you submit a performance review through your company's system, it becomes part of an employee's official personnel file. That file can be subpoenaed. It can be submitted as evidence in an EEOC complaint. It can be used to support or contradict a wrongful termination claim. It can appear in arbitration, mediation, or open court.

This does not mean you need to write like a lawyer. It means you need to write with the awareness that your words have weight beyond the conversation you're having today. The standard is not perfection. The standard is this: could a reasonable, objective third party read this review and understand exactly what the employee did, how it was evaluated, and why?

If the answer is yes, you are almost certainly fine. If the answer is no, meaning your review is vague, emotionally charged, inconsistent with the rating, or contains language that references protected characteristics, then you have a problem that needs to be fixed before that document is signed.

## Where HR Earns Its Stripes

In many organizations, Human Resources sits downstream in the review process, collecting completed forms, filing them away, and moving on. That is a missed opportunity for everyone involved, including the manager.

In well-run organizations, HR plays an active audit role before reviews are finalized. They read the language. They flag the inconsistencies. They ask the uncomfortable questions that protect both the employee and the company. This is not HR being difficult. This is HR doing its job at the highest level.

The best HR professionals understand that a poorly written review is not just a legal risk; it's a leadership failure. When a manager writes something vague, biased, or indefensible, it tells a story about that manager's judgment, not just their writing. HR's job is to catch that before it becomes permanent.

## The HR Audit: What They're Looking For

When an experienced HR professional audits a performance review, they are running through a mental checklist. Here is that checklist made explicit, both so you understand what they're doing and so you can apply it yourself before you submit.

| HR Audit Question | What a Well-Written Review Provides |
|---|---|
| **Does the rating match the written narrative?** | A glowing written review paired with a Meets Expectations rating, or a critical narrative paired with an Exceeds, creates legal inconsistency. Employees and their attorneys notice this immediately. |
| **Are all statements tied to observable behavior or measurable results?** | Opinion and adjective-heavy reviews ('bad attitude,' 'difficult personality') are difficult to defend in any formal process. Every claim should trace back to something that happened, was observed, or was measured. |
| **Does any language reference a protected characteristic, whether directly or indirectly?** | This includes obvious references (age, race, religion) and indirect ones ('not a cultural fit,' 'doesn't communicate the way we expect,' 'hard to relate to'). These phrases can trigger discrimination claims. |
| **Is the level of documentation consistent with similar employees at the same performance level?** | If one employee receives three pages of detailed developmental feedback and another at the same rating level receives two sentences, the disparity itself can become evidence of differential treatment. |
| **Does the review reflect only the current rating period?** | Bringing up incidents from prior rating periods that were already documented and resolved is legally problematic. The review should reflect this year, not a history lesson. |
| **If this review were used to support a termination, would it hold up?** | HR thinks backward from worst-case scenarios. If the review doesn't clearly document a performance gap, it cannot be used to support discipline or separation, leaving the organization exposed. |
| **Has the manager documented the same issue in prior reviews?** | Consistency over time is critical in employment law. A performance problem that appears for the first time in a termination-year review, with no prior documentation, is nearly impossible to defend. |

If you have followed the documentation principles in this book, then you will pass most of this audit without incident. The two areas where good-faith managers most commonly run into trouble are the language of protected characteristics, which is often unintentional, and the problem of inconsistency between their written comments and their rating. Both are fixable. But you have to know they exist.

> *A note on the manager-HR relationship: HR is not your adversary in this process. They are your quality control. The manager who builds a genuine partnership with HR, asking questions before reviews are due, inviting feedback on draft language, and treating the audit as a professional service rather than a bureaucratic obstacle, is the manager who almost never ends up in a legal dispute over a review. That partnership is worth investing in.*

## The Phrases That Will Get You in Trouble

What follows is not a list of things only bad managers write. These are phrases that appear regularly in performance reviews written by people who care about their employees and believe they are doing the right thing. The problem is not intent. The problem is impact.

Each phrase below carries legal risk, creates ambiguity, or both. The table shows you what makes it dangerous and, more usefully, what to write instead.

## Category 1: Language That References Protected Characteristics

These are the highest-risk phrases in any review. Some reference protected characteristics directly. Others use language that courts have repeatedly found to be coded references to age, race, national origin, disability, or other protected classes. The intent behind the phrase is irrelevant; the words themselves create exposure.

| The Phrase | Why It's Legally Risky | Use This Instead |
|---|---|---|
| "Doesn't communicate the way we expect" | Can be read as bias against non-native speakers, cultural communication styles, or employees with disabilities affecting communication. | "Written and verbal communication does not consistently meet the standards for this role. Specifically: [name the gap]." |
| "Not a cultural fit" | Widely recognized as coded language that can mask bias based on race, national origin, religion, or age. Almost never defensible. | "The candidate/employee has not demonstrated alignment with our core operating principles of [name them specifically]." |
| "Seems tired / low energy lately" | Can imply age bias (40+) or reference a disability or medical condition, both of which are protected. Appearance-based observations belong nowhere in a review. | "Productivity and engagement have declined this period. Specifically: [name metrics or observed behaviors]." |
| "Hard to relate to / hard to connect with" | Vague and potentially biased. Courts have found this language to reflect cultural or racial bias in numerous cases. | "Has not built the cross-functional relationships required at this level. Specific examples: [name them]." |
| "Since becoming a mother/father..." | Direct reference to parental status, which is protected under federal and many state laws. Pregnancy and family status are off-limits entirely. | Simply do not reference family circumstances. Document the work behavior or result only. |
| "At this stage of his/her career..." | Strongly implies age. Courts have used this phrase as evidence of age discrimination (ADEA applies to employees 40+). | "At this level of the organization..." or name specific experience gaps without referencing tenure or age. |
| "English could be stronger" | National origin is a protected characteristic. Language proficiency should only be noted when it directly and demonstrably affects job performance. | "Written communication in client-facing reports does not consistently meet expectations. Specific examples: [name them]." |

## Category 2: Vague Language That Is Legally Indefensible

This category is less dramatic than the first but causes more day-to-day problems. Vague language, the kind that sounds meaningful but says nothing specific, is indefensible in two critical ways. First, it gives the employee no actionable information. Second, if the review is ever used to support a disciplinary action or termination, it provides no foundation. 'Needs to improve attitude' cannot support a PIP. 'Showed consistent quality issues' without specifics cannot support a separation.

| The Phrase | Why It's Legally Risky | Use This Instead |
|---|---|---|
| "Bad attitude" | Pure opinion with limited legal standing. Unobservable, unmeasurable, and very difficult to defend. Employees have successfully challenged discipline based on this language alone. | "On [specific dates/occasions], responded to correction by [describe specific behavior]. This falls below the professional standard expected at this level." |
| "Doesn't take initiative" | Vague and subjective. What does initiative mean in this role? Without a definition and examples, the phrase is meaningless and legally empty. | "Has not proactively identified process improvement opportunities in [area]. By contrast, peers in equivalent roles have [specific example]." |
| "Needs to improve communication" | Every single review in every organization contains some version of this sentence. It is nearly meaningless without specifics and creates no legal record. | "Written updates to the team on project status have been inconsistent, resulting in [specific impact]. This needs to improve by [specific expectation]." |
| "Does not meet expectations" (with no elaboration) | A rating without written support is not documentation; it is an assertion. An attorney for the employee will make that distinction very clearly. | Follow every below-expectation rating with at least two to three specific examples of the gap, named behaviors or results, and clear statements of what was expected. |
| "Always" or "Never" | Absolute language is almost always factually inaccurate and will be challenged. One counterexample defeats the entire statement. | Use frequency language: "On multiple occasions," "Consistently below the standard," "Rarely met the deadline in this area." |
| "Has a lot of potential" | Sounds positive. Is actually dangerous. This phrase has been used in wrongful termination cases as evidence that the employee should not have been let go. | "Has demonstrated growth in [area]. To advance to the next level, must demonstrate [specific competency] with consistency over the next rating period." |
| "Everyone knows that..." | Implies the employee should have known something that was never formally communicated. Creates the impression of an unfair standard. | "The expectation, as communicated on [date] and documented in [file/email/meeting notes], was [specific standard]." |

**Kevin Baker**

## Category 3: Language That Contradicts the Rating

This category is the one most managers don't see coming. The problem isn't any single phrase. It's the relationship between what you wrote and what you scored. When those two things don't match, the review creates its own contradiction, and that contradiction can be used against the organization.

| The Phrase | Why It's Legally Risky | Use This Instead |
| --- | --- | --- |
| Exceeds rating with no written evidence of exceptional performance | If the review is ever challenged, whether by the employee's peers, in a calibration dispute, or in legal discovery, you need to be able to point to specific behaviors that justify an above-standard rating. | Every Exceeds rating needs at least two specific examples of performance that went beyond what was required, with measurable impact named. |
| Glowing written narrative with a Meets or below rating | Employees and their attorneys read narrative first, rating second. If the narrative reads like an Exceeds and the rating is Meets, you will be formally asked to explain the gap. | If the rating is Meets, the narrative should honestly reflect both what was done well and what still has room to grow. The balance should reflect the rating. |
| Critical written feedback with no mention in the rating summary | If a serious performance gap is described in the body of the review but not reflected in the overall rating, it signals to HR, and later to attorneys, that the gap was not taken seriously. | Critical performance issues must be reflected in the rating. A significant gap in one area should pull the overall rating down, or you must explain in writing why it did not. |
| First mention of a serious problem at termination | If an employee is terminated for a performance issue that does not appear in any prior review, the organization has almost no legal ground to stand on. | Document early, document often, document specifically. Every performance conversation of significance should have a paper trail. The review is the summary, not the only record. |

## Category 4: The 'Kind' Phrases That Create Problems

These are the hardest to give up, because they come from a good place. A manager who wants to protect an employee from discouragement, who doesn't want to close a door, who believes the person will turn it around. That manager writes phrases like these with genuine care. But the impact of the language is the same regardless of the intent.

| The Phrase | Why It's Legally Risky | Use This Instead |
| --- | --- | --- |
| "I'm sure things will improve" | Implies the manager accepts the current performance gap as temporary without documented improvement conditions. Can be used to argue that discipline was not warranted. | "Improvement in [specific area] is expected by [specific timeframe]. Progress will be reviewed at the mid-year checkpoint." |
| "This was a tough year for everyone" | Contextualizes poor performance in a way that could be read as excusing it. Context belongs in a review, but it should not be used to explain away gaps in performance. | "External conditions this year created significant operational challenges. Despite those conditions, the expectation was [X]. The result was [Y]." |
| "I know you've been going through a lot personally" | References personal circumstances that may include protected characteristics (medical condition, family status). Keep personal context entirely out of performance documentation. | Document the work performance only. If accommodations were made, document them separately through the appropriate HR process, not in a performance review. |
| "We'll work on this together" | Implies the manager shares responsibility for the employee's performance gap. Management support is appropriate; shared accountability for individual performance results is not. | "I will provide [specific support, coaching, resources]. The accountability for improvement in this area belongs to the associate." |
| "You're one of my favorites" | Legal risk on two fronts: favoritism claims from other employees, and an implied promise of continued employment or positive treatment. | Simply do not write it. Recognition of value belongs in the review, framed around performance rather than personal preference. |

**Kevin Baker**

# The Pre-Submission Self-Audit: Seven Questions

Before you submit any review, run through these seven questions. If you can answer yes to all of them, you are in good shape. If any give you pause, revise before you submit. It is significantly easier to fix language before the document is signed than after.

- **Does every claim in this review trace back to an observable behavior or a measurable result?**

If a statement exists in your review that cannot be supported by something you observed, a metric you measured, or a documented event, remove it or replace it with something that can.

- **Does the written narrative match the rating I assigned?**

Read the narrative first, then the rating. Ask yourself: if a stranger read this narrative, would they predict this rating? If not, one of them needs to change.

- **Does any word or phrase in this review reference, whether directly or indirectly, a protected characteristic?**

Age, race, gender, religion, national origin, disability, pregnancy, family status. Read every sentence with this lens. If anything gives you pause, take it to HR before submitting.

- **Am I bringing up issues from prior rating periods that have already been addressed?**

If a performance issue was documented, discussed, and resolved in a prior cycle, it does not belong in this review. The current review covers the current period. Prior resolved issues are closed.

- **Have I been consistent across similar employees?**

If two employees have similar roles and similar performance levels, their reviews should reflect similar standards of documentation and evaluation. Significant disparity in detail or tone is a legal exposure point.

- **If this review were used to support a termination tomorrow, would it hold up?**

Think worst-case. If this employee's performance does not improve and a separation becomes necessary, does this review document the gap clearly enough to support that decision? If not, document more specifically now.

- **Would I be comfortable if HR, senior leadership, or legal counsel read every word of this review right now?**

This is the gut-check question. If the answer is anything other than yes, keep editing. The discomfort you feel reading it now is significantly less than the discomfort of defending it later.

# How to Build a Real Partnership with HR

If you take nothing else from this chapter, take this: the managers who have the fewest legal problems with performance reviews are not the ones who know the most employment law. They are the ones who have built a genuine working relationship with their HR partner.

Here is what that relationship looks like in practice, as well as what it does not look like.

## What the Partnership Looks Like

- You send HR draft language before reviews are finalized, not to get approval, but to get a second set of eyes on anything you're uncertain about.

- When HR flags something in your review, your first response is curiosity, not defensiveness. 'Help me understand what the concern is' is the right question.

- You ask HR at the start of review season: 'What are you seeing across the organization this year that I should be aware of as I write?' They know things you don't.

- When a performance situation is developing, before it reaches the review, loop in HR early. They can help you document correctly from the start.

- You treat the HR audit as a professional service, not an administrative hurdle. A good auditor is protecting you as much as they're protecting the organization.

## What the Partnership Does Not Look Like

- Submitting reviews at the last minute to avoid HR review time.

- Arguing with HR feedback because you believe you know better. (You may. But you may also be wrong. And if you're wrong, you're the one exposed.)

- Treating HR as the enemy when they flag your language. The flag is the service.

- Assuming that because a review passed HR once, similar language is always safe. Employment law evolves. So do organizational standards.

- Waiting for the termination conversation to start documenting. By then, it's too late to build the record you need.

The best HR professionals I worked with over two decades in distribution operations had the same quality: they were direct. They didn't soften the feedback when a review had a problem. They named it, explained it, and helped fix it. That directness, the same directness this book advocates for in your performance reviews, is what made them genuinely valuable partners.

If your HR partner operates that way, treat them like gold. If they don't, advocate for a more substantive review process within your organization. The legal and human cost of getting this wrong is too high to leave it to chance.

## Chapter 8 — Key Takeaways

- Every performance review is a legal document from the moment it is submitted.
- The most dangerous language is often unintentional, consisting of references to protected characteristics hidden in everyday phrases.
- Vague language is not safe language. Specificity protects everyone.
- The narrative and the rating must tell the same story. When they don't, the inconsistency becomes the story.
- HR auditing performance review language is HR at its best. Build that partnership before you need it.
- Run the seven-question self-audit on every review before you submit. Every time.

# Chapter 9

# The Full Review Template

*A template is not a shortcut. It is a scaffold. The words still have to be yours: specific to your employee, grounded in what you observed, honest about what you saw. What the template does is remove the blank page. And for most managers, the blank page is the hardest part.*

The language framework, the phrase banks, the performance-level writing guides, and the legal guardrails developed throughout this book all come together here, into a single guided template you can use for every review you write, from the first one in your career to the hundredth.

This chapter has three parts. First, the blank template with guided prompts, the document you will actually print or adapt for your own use. Second, three completed sample reviews at different performance levels, drawn from real operational leadership experience and anonymized for privacy. Third, a brief guide on how to adapt this framework if your organization uses a competency-based review system, a goals-based system, or a hybrid of both.

Before you use the template, one reminder that runs throughout this book: the review should never be the first time an employee hears any of this. Everything in these pages should reflect conversations, coaching moments, and feedback that happened throughout the year. If you are delivering a surprise in a performance review, something went wrong long before review season arrived.

## Part One: The Full Review Template with Guided Prompts

Complete each section in order. The prompts are designed to draw out specific, evidence-based language. If a prompt feels difficult to answer, that is usually a signal that you need more notes, not that the employee doesn't deserve a full review. Go back to your documentation from the year before you start writing.

## Section 1 — Employee Information

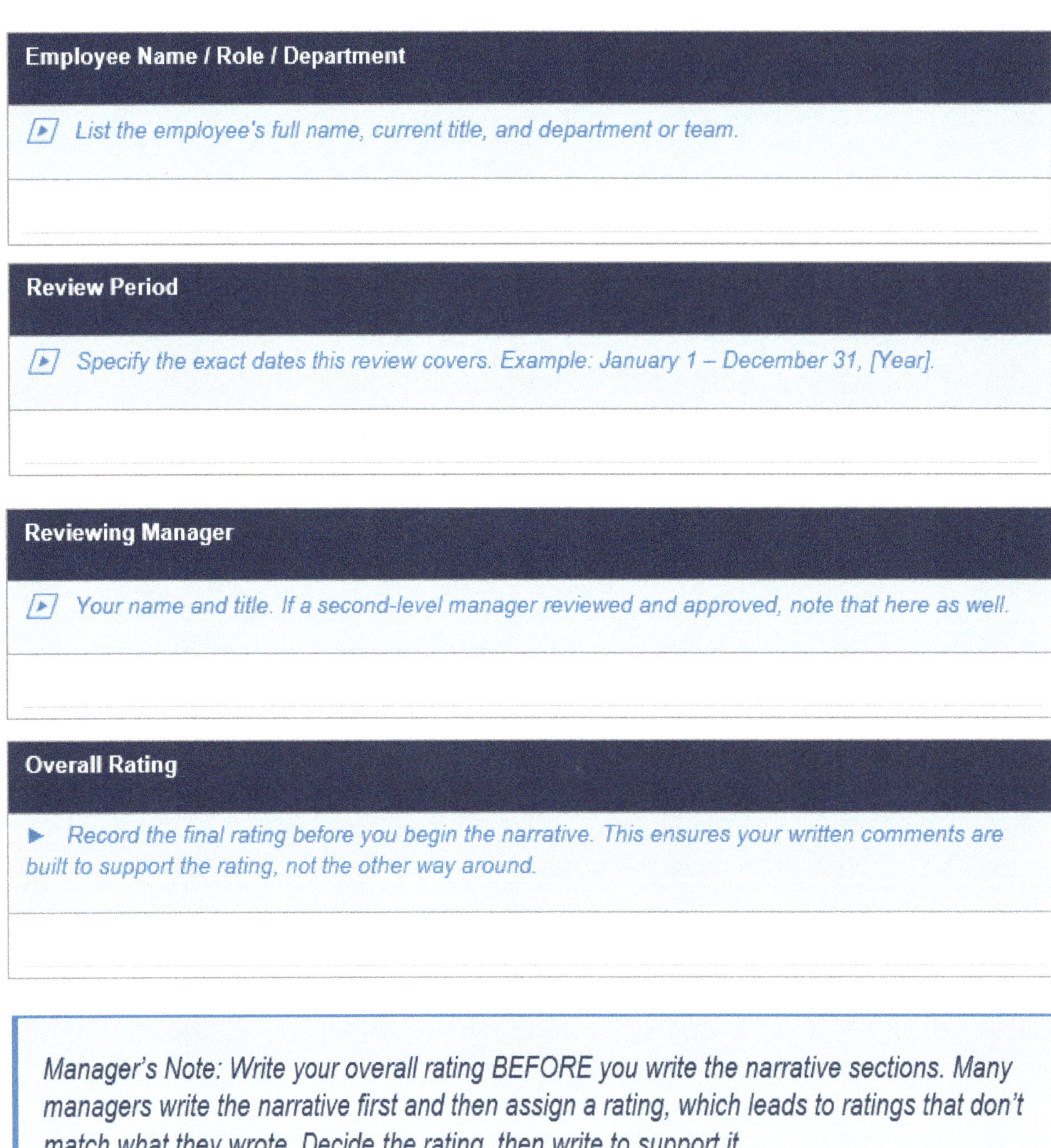

**Employee Name / Role / Department**

▶ List the employee's full name, current title, and department or team.

**Review Period**

▶ Specify the exact dates this review covers. Example: January 1 – December 31, [Year].

**Reviewing Manager**

▶ Your name and title. If a second-level manager reviewed and approved, note that here as well.

**Overall Rating**

▶ Record the final rating before you begin the narrative. This ensures your written comments are built to support the rating, not the other way around.

Manager's Note: Write your overall rating BEFORE you write the narrative sections. Many managers write the narrative first and then assign a rating, which leads to ratings that don't match what they wrote. Decide the rating, then write to support it.

## Section 2 — The Year in Review

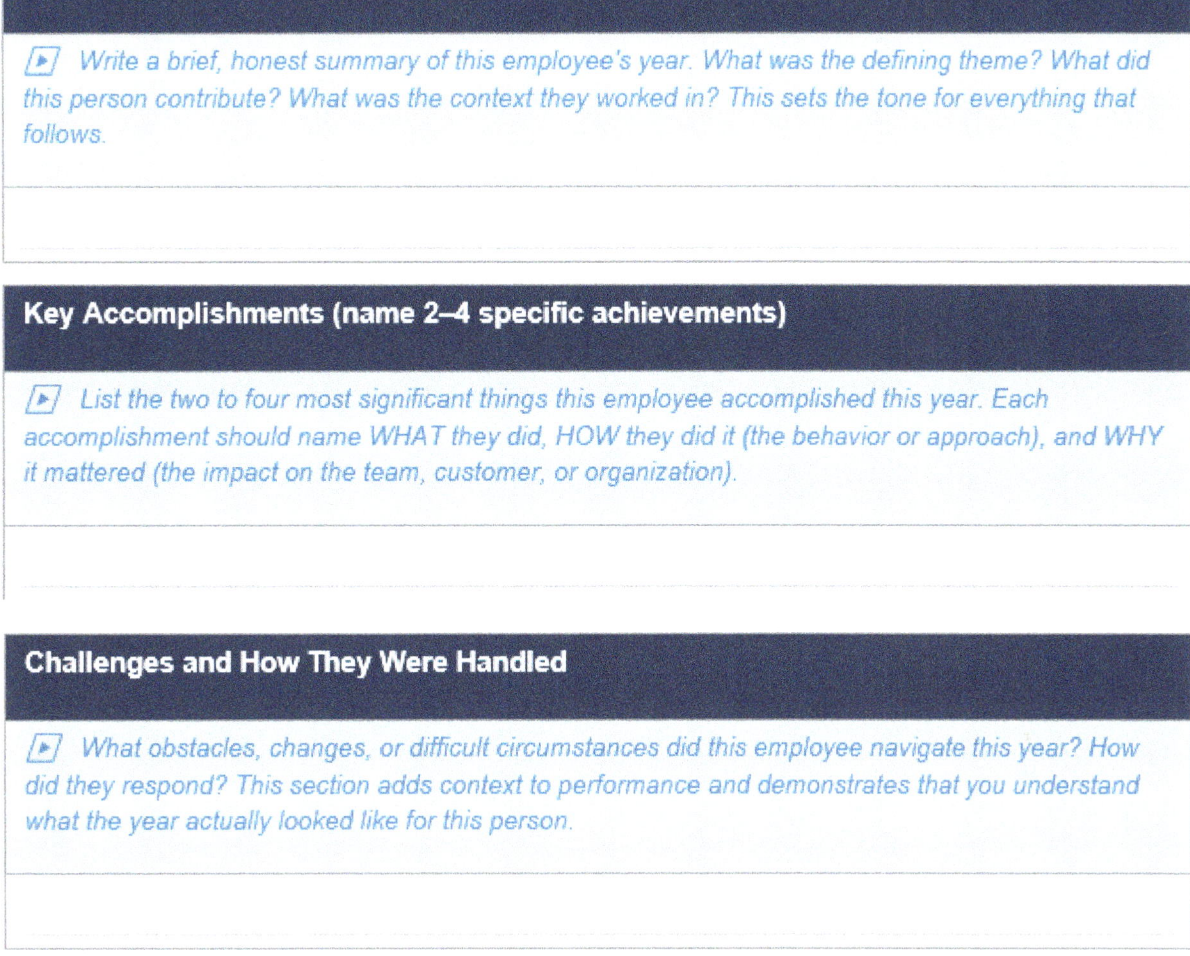

**Opening Summary (2–4 sentences)**

▶ *Write a brief, honest summary of this employee's year. What was the defining theme? What did this person contribute? What was the context they worked in? This sets the tone for everything that follows.*

**Key Accomplishments (name 2–4 specific achievements)**

▶ *List the two to four most significant things this employee accomplished this year. Each accomplishment should name WHAT they did, HOW they did it (the behavior or approach), and WHY it mattered (the impact on the team, customer, or organization).*

**Challenges and How They Were Handled**

▶ *What obstacles, changes, or difficult circumstances did this employee navigate this year? How did they respond? This section adds context to performance and demonstrates that you understand what the year actually looked like for this person.*

## Section 3 — Performance Area Ratings and Written Comments

For each performance area below, assign a rating AND write a substantive comment. A rating without a written comment is not documentation; it is an assertion. Every rating must be supported by specific language that explains what you observed and why you rated it as you did.

Use your organization's rating scale. Common scales include:

- Exceeds Expectations / Meets Expectations / Does Not Meet Expectations (3-point)
- Exceptional / High Meets / Meets / Improvement Needed / Unsatisfactory (5-point)
- Outstanding / Exceeds / Meets / Below / Unsatisfactory (5-point variant)

The labels matter less than the consistency. Whatever scale your organization uses, apply it the same way for every employee at the same level.

### Performance Area 1: Leadership

▶ *How does this employee lead, formally or informally? Do they influence others, build trust, remove barriers, and model the behaviors they expect from their team? Name specific examples. Note both what is working and what still has room to grow.*

### Performance Area 2: Accountability

▶ *Does this employee take genuine ownership of their responsibilities and results? When things go wrong, how do they respond? Are their deliverables consistent and timely? Name the evidence, both positive and developmental.*

### Performance Area 3: Quality of Work / Excellence

▶ *Is the work this employee produces accurate, thorough, and delivered on time? What does the quality of their output signal about their standards? Cite specific metrics where available.*

**Kevin Baker**

## Performance Area 4: Communication

*▶ How does this employee communicate, up, down, and across the organization? Are they clear? Do they listen? Do they share information proactively? Note both written and verbal communication if both are relevant to the role.*

## Performance Area 5: Teamwork and Collaboration

*▶ How does this employee contribute to the team beyond their individual responsibilities? Do they support peers, share knowledge, and help build the collective? Are there cross-functional relationships to note?*

## Performance Area 6: [Role-Specific Competency]

*▶ Insert the competency most relevant to this specific role: safety, customer focus, technical skill, financial management, project execution, or others as defined by your organization. Name the standard and assess against it specifically.*

## Section 4 — Goals Review

For each goal established at the start of the review period, record the original goal, the result, and your written assessment. If no formal goals were set, use this section to document what the key expectations were and how they were met.

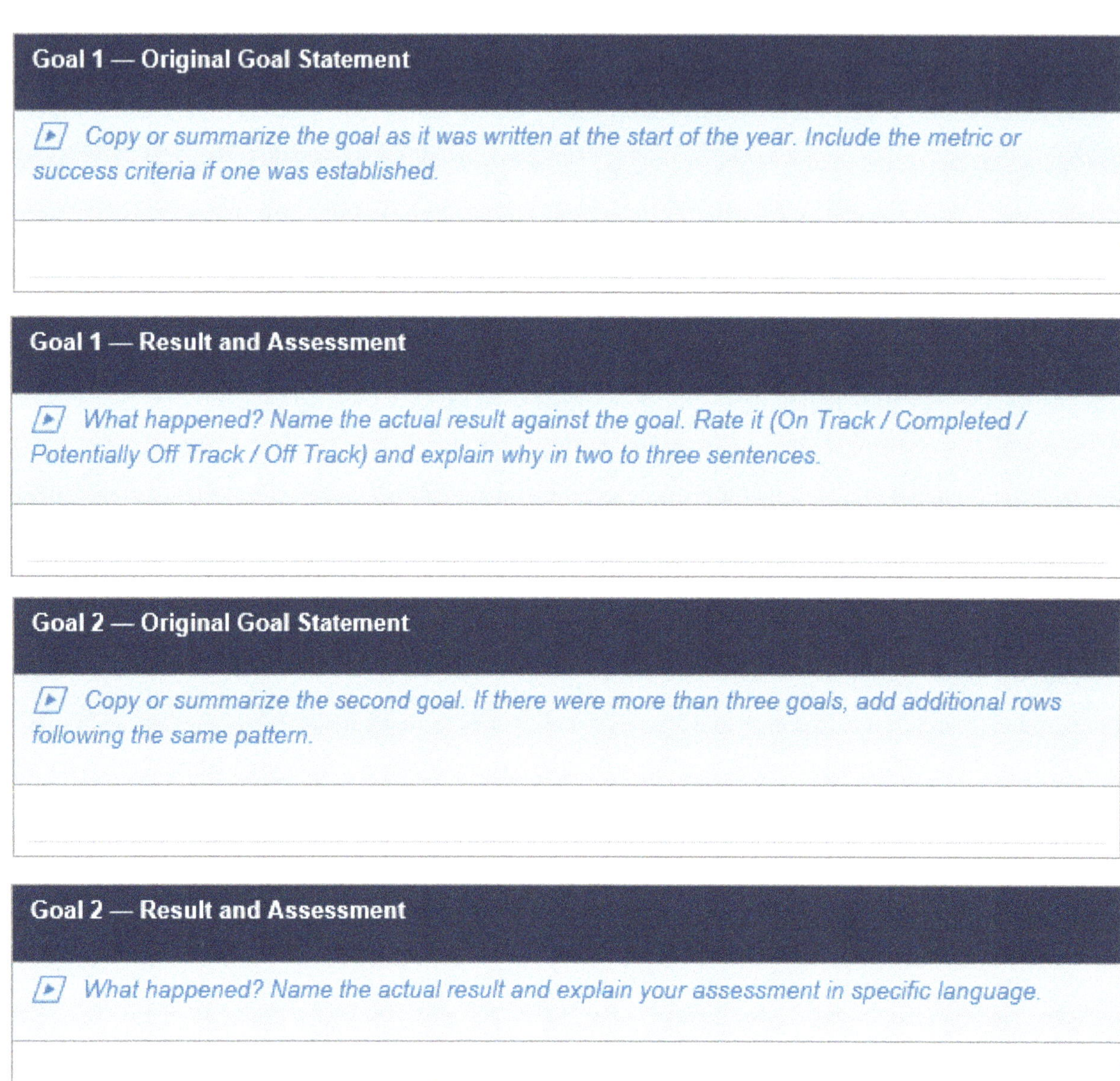

**Goal 1 — Original Goal Statement**

*▶ Copy or summarize the goal as it was written at the start of the year. Include the metric or success criteria if one was established.*

**Goal 1 — Result and Assessment**

*▶ What happened? Name the actual result against the goal. Rate it (On Track / Completed / Potentially Off Track / Off Track) and explain why in two to three sentences.*

**Goal 2 — Original Goal Statement**

*▶ Copy or summarize the second goal. If there were more than three goals, add additional rows following the same pattern.*

**Goal 2 — Result and Assessment**

*▶ What happened? Name the actual result and explain your assessment in specific language.*

## Goal 3 — Original Goal Statement

▶ *Copy or summarize the third goal.*

## Goal 3 — Result and Assessment

▶ *What happened? Name the actual result and explain your assessment in specific language.*

## Section 5 — Development and Growth

### Developmental Focus Area(s) for the Coming Year

▶ *Name one to two specific areas where this employee has the most opportunity to grow. Be specific about what 'growth' looks like: what would you see or measure that would tell you they have improved in this area? Frame this as an extension of their existing strengths where possible.*

### Recommended Development Activities

▶ *What specific activities, experiences, courses, or relationships would support this employee's growth? Be concrete: not 'seek out leadership opportunities' but 'lead the Q3 process improvement project and present findings at the manager summit.' Vague development plans are not development plans.*

### Career Path Discussion

▶ *What did you and this employee discuss about their longer-term direction? What do they want? What do you see as realistic? What is the next step, and what would it take to get there? This section may be brief but it signals that the conversation happened.*

Kevin Baker

## Section 6 — Closing Comments and Next Steps

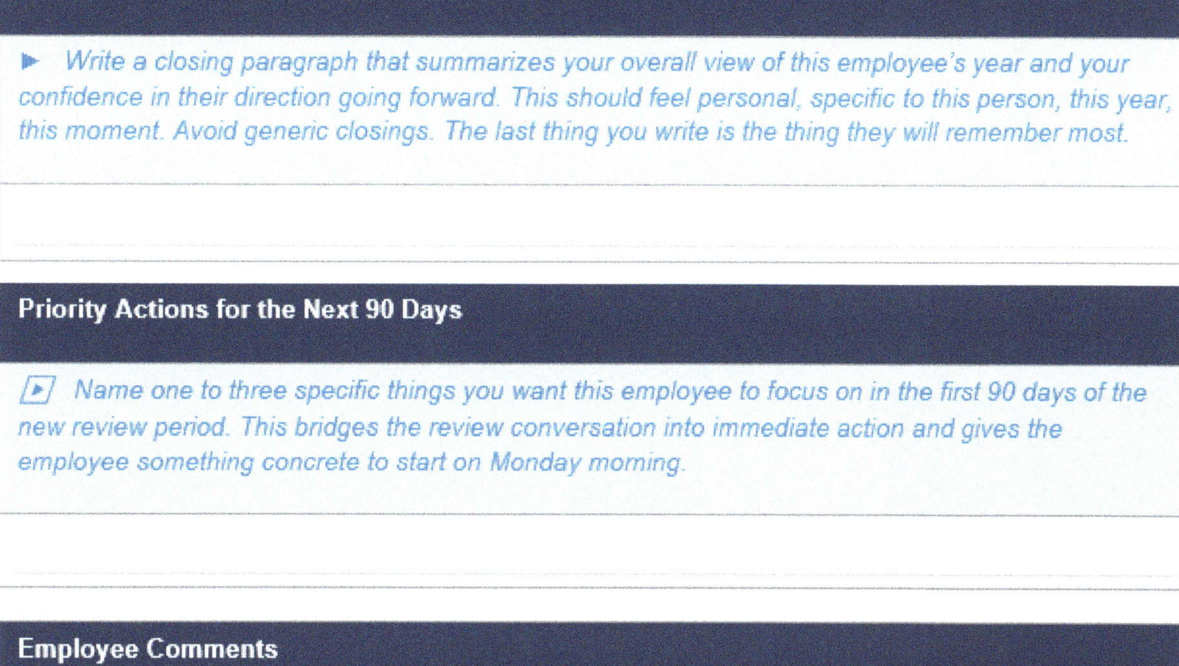

### Manager's Closing Statement

▶ *Write a closing paragraph that summarizes your overall view of this employee's year and your confidence in their direction going forward. This should feel personal, specific to this person, this year, this moment. Avoid generic closings. The last thing you write is the thing they will remember most.*

### Priority Actions for the Next 90 Days

▶ *Name one to three specific things you want this employee to focus on in the first 90 days of the new review period. This bridges the review conversation into immediate action and gives the employee something concrete to start on Monday morning.*

### Employee Comments

▶ *Leave this section blank for the employee to complete. Prompt them to share their perspective on the year, any context you may not have had, and their own development goals. Their comments become part of the official record.*

# Part Two: Three Completed Sample Reviews

The following three reviews demonstrate how the template above translates into finished, professional review language at three distinct performance levels. Each is drawn from real operational leadership experience and anonymized to protect the individuals involved. Names have been replaced with role and location identifiers.

Read all three, not just the one that matches your current situation. Understanding the full range is what makes you a more precise writer at every level.

## Sample Review A — Exceptional Performance

| | |
|---|---|
| **Employee** | DC Manager, Illinois |
| **Role** | Distribution Center Manager |
| **Review Period** | January 1 – December 31 |
| **Overall Rating** | High Meets / Exceptional (LEAP Framework) |

## Opening Summary

This was an extraordinary year under extraordinary circumstances. Stepping back into the role after personal leave and directly into the operational challenges of a global pandemic, this manager did not miss a step. While the broader organization navigated anxiety, uncertainty, and disruption, the Illinois distribution center became the stable center of gravity that everyone else leaned on. That stability was not accidental. It was built, deliberately and daily, through exceptional leadership.

## Leadership — Exceptional / Exceeds Expectations

You stepped back into leadership right in the middle of a pandemic. The organization as a whole was anxious and in disarray. You returned and brought with you a soothing voice and a stabilizing presence. The distribution center became the rock of the organization, remaining open and operating after everything else had closed. It became the bright light at the end of a dark tunnel for many during these turbulent times.

You accomplished this through effective leadership: clear and consistent communication with your team, empathetic listening and decision-making, and by modeling stability and calmness when the pressure to panic was real. The building represents the life of this organization, and you protected it, and the people in it, with your leadership. There is no higher mark in this competency.

## Accountability — Exceptional / Exceeds Expectations

Accountability in this context meant making decisions quickly that other leaders were still deliberating over. Enhanced safety protocols were not suggested; they were implemented, communicated, and enforced consistently. When associates needed answers, you had them. When the organization needed a model for how to operate during a crisis, your building was the model.

The associate engagement survey results from this period are a direct reflection of the environment you created, one where people felt safe, informed, and valued even when the world outside the building felt uncertain. That result does not happen without a leader who owns every aspect of the environment they are responsible for.

## Performance — Meets Expectations

In many ways, overall performance metrics were put on hold this year. Shrinking volumes can only be partially offset by labor reductions. In this environment, you pulled every lever available to you, managed expenses with discipline, and kept your associates engaged through a period that would have unraveled a less experienced team.

As we exit this difficult period, rely on your team to manage much of the day-to-day and focus on what can be accomplished creatively now and how to lead this building to new heights as conditions normalize. The foundation is solid. Build on it.

## Closing Statement

Continue to be the rock others feed off of and gravitate toward. It can be a tremendous weight to bear, but you have the wisdom, the emotional intelligence, and the leadership command to carry it. This building has a psychological effect on everyone around it. It represents normalcy. It represents company energy. It represents what this organization is capable of at its best. Continue growing and nurturing it, and the people within it.

**Kevin Baker**

## Sample Review B — Solid Contributor / Building Momentum

| | |
|---|---|
| **Employee** | DC Manager, Texas |
| **Role** | Distribution Center Manager |
| **Review Period** | January 1 – December 31 (First Full Year) |
| **Overall Rating** | Meets Expectations |

## Opening Summary

This was a first full year in a new role, a new organization, and a new market, and by every reasonable measure, it was a successful one. This manager arrived mid-year in the prior cycle and spent the first half of this year building the credibility, relationships, and operational command that the role required. By the second half, the results were visible in the metrics, in the team, and in the trust his peers placed in him. Dallas is now fully his. The opportunity ahead is significant.

## Communication — Exceeds Expectations

This was the standout competency of the year. The ability to articulate complex operational thinking, both in writing and in conversation, distinguished this manager early and continued to build his credibility throughout the year. Peers sought him out not because of his title but because of the quality of his thinking and the clarity with which he expressed it. He listened as often as he spoke. That balance, rare among strong-minded operational leaders, is what made him a trusted voice in a peer group of experienced managers.

## Leadership — Meets Expectations

You are able to influence others and gain their genuine commitment rather than their reluctant compliance. This comes from being a leader who asks for input and actually uses it, making team members feel valued as experts in their own positions. Your inventory specialist and your second-shift supervisor are both examples of people who have grown visibly under the confidence and autonomy you have extended to them.

There was a period mid-year where you struggled to find your leadership footing. The second half of the year, you hit your stride, and the metrics reflect it clearly. The remaining opportunity is knowing when to shift from democratic to autocratic, when to stop asking and start deciding. Do not let your own self-created barriers limit what this building can become.

**Kevin Baker**

## Work Quality and Productivity — Meets Expectations

This is the area with the most room for growth in the coming year. Productivity and quality metrics in the Texas facility have not yet reached the level expected for a building of this size and scope. The standard for next year is clear: 52 cartons per hour processed, sustained across the full year. For the last several months, this target has not been reached with regularity.

As the leader, you set the standard. Make the job easier for your group by driving process improvements, holding underperformers accountable, and refusing to fall into the excuse trap. The data is there. The team is capable. What is needed now is sustained execution and an unwillingness to accept less.

## Closing Statement

This has been a genuinely impressive start. I have been thoroughly impressed by your attention to detail, your ability to build rapport across the organization, and your willingness to listen, learn, and act. You are already viewed as an asset by your peers and by me. The coming year belongs to you. Own the building, own your leadership team, and proudly own the results of your dedication. Dallas is ready for its next chapter. Lead it there.

## Sample Review C — Performance Requiring Focused Improvement

The following sample demonstrates how to write an honest, specific, and professionally sound review for an employee whose performance has not met expectations in one or more critical areas. This type of review is the hardest to write well and the most important to get right. Notice that the tone is direct without being harsh, the gaps are named specifically, and the path forward is clear.

| | |
|---|---|
| **Employee** | DC Supervisor, Illinois |
| **Role** | Distribution Supervisor (Hourly Team Lead) |
| **Review Period** | January 1 – December 31 |
| **Overall Rating** | Improvement Needed / Low Meets Expectations |

## Opening Summary

This was a year of mixed results. There are genuine areas of contribution that deserve recognition: this supervisor's relationships with associates are strong, and the willingness to show up and work hard is not in question. What is in question is whether the performance, consistency, and ownership expected at the supervisor level have been demonstrated with the frequency and reliability the role demands. This review addresses both the contribution and the gap with equal honesty.

## Reliability and Consistency — Improvement Needed

Attendance and schedule adherence did not meet the standard expected for a supervisor role during this review period. On seven occasions, the supervisor was late to shift or absent without adequate advance notice. The impact on the team was direct: associates began shifts without clear direction, and other supervisors absorbed additional coverage responsibility. This pattern was discussed on three separate occasions during the year and has not shown sustained improvement.

The expectation going forward is clear: supervisors are expected to be present, on time, and prepared at the start of every shift. Any absence requires advance notification except in genuine emergencies. This is a non-negotiable standard of the role and must be met consistently in the coming year.

## Accountability — Improvement Needed

When errors occurred on this supervisor's shift during this period, the response was frequently to explain the circumstances rather than to own the outcome. Accountability does not mean taking blame for events beyond your control; it means owning the environment your team operates in and the standards they are held to. On multiple occasions this year, quality errors were traced back to process steps that had not been reinforced consistently by shift supervision.

The expectation is that this supervisor will move from explanation to action. When something goes wrong on your shift, the first question should be: what do I need to change to prevent this from happening again? That question, asked consistently, is the foundation of supervisory accountability.

## Associate Relationships — Meets Expectations

This is a genuine strength and should be named as one. Associates respond well to this supervisor, feel comfortable raising concerns, and describe the working environment on this shift positively. That relational credibility is real and it matters. The opportunity is to leverage that trust in service of higher performance standards and become the kind of leader whose team works hard not just because they like the supervisor, but because the supervisor holds them to something worth being proud of.

## Development Focus and Path Forward

The coming year will function as a critical evaluation period. The two areas identified above, reliability and accountability, are not optional growth targets. They are requirements of the role. Failure to demonstrate sustained improvement in both areas by mid-year will result in a formal performance improvement plan.

The support available to make this work: bi-weekly one-on-one check-ins with the managing director, a formal mentorship connection with a senior supervisor in the network, and access to the LearnOn supervisory development curriculum. The resources are in place. The accountability for using them and demonstrating improvement belongs entirely to the associate.

## Closing Statement

This review is written with directness because directness is the most respectful thing I can offer. The gaps named here are real. They are also fixable. The associate relationships, the work ethic, and the desire to do well are all present. What is needed now is the consistency and ownership that translate those qualities into supervisory performance. I am invested in seeing that happen. The path forward is clear.

# Part Three: Adapting This Framework to Your Organization's System

Every organization structures its performance review differently. Some use competency frameworks. Some use goal-based systems. Some use a hybrid. Whatever system your organization uses, the principles in this template apply. Here is how to adapt them.

## If Your Organization Uses a Competency-Based System

### The LEAP Framework: A Brief Explanation

American Hotel Register Company used a competency framework called LEAP (Leadership, Excellence, Accountability, and Performance) as the backbone of its annual review process. LEAP defined not just what employees were expected to accomplish, but how they were expected to show up: the behaviors, values, and professional qualities the organization held in highest regard.

Each of the four competencies was rated independently, with its own narrative comment. A manager writing a LEAP-based review was not writing one unified evaluation; they were writing four mini-reviews, each anchored to a defined standard. Leadership meant inspiring confidence in others and setting an example under pressure. Excellence meant consistent quality and a bias for getting things right the first time. Accountability meant owning outcomes, the good ones and the ones that fell short. Performance meant delivering the operational results the role required.

If your organization uses a competency framework, by any name and with any number of dimensions, the writing guidance throughout this book applies directly. Simply substitute your organization's competency names wherever you see the LEAP categories referenced. The structure of the writing is identical. What changes is only the label.

## If Your Organization Uses a Goals-Based System

Goals-based systems rate employees primarily on whether they achieved the objectives established at the start of the year. Section 4 of this template, the Goals Review, becomes the central section rather than a supplemental one. For each goal, name the original expectation, the actual result, and your written assessment of why the result occurred. Even in a goals-only system, the written narrative around each goal is what separates a useful review from a checkbox exercise.

## If Your Organization Uses a Hybrid System

Hybrid systems, which weight both goals and competencies, are the most common in mid-to-large organizations. The template in this chapter is designed for hybrid use. Adjust the weighting language to match your organization's split. The most common configurations are 50/50 goals and competencies, or 60/40 with goals carrying more weight. Whatever the split, the quality of the written language in both sections determines how useful the review actually is.

## Chapter 9 — Key Takeaways

• Write your overall rating before you write your narrative, not after.

• Every performance area rating must be supported by specific written language. A rating without a comment is an assertion, not documentation.

• The blank template in Part One works for any review system, whether competency-based, goals-based, or hybrid.

• The three sample reviews demonstrate the full range: exceptional performance, solid development, and honest improvement feedback.

• The closing statement is the thing they will remember. Write it like it matters, because it does.

• The form is the record. The conversation is the review. Do both well.

# Chapter 10

# The Mid-Year Check-In

*The annual review should never be the first time an employee hears where they stand. If it is, you have already failed them. Not at review time, but six months earlier, when the conversation that could have changed the outcome never happened.*

The mid-year check-in is the most underused tool in a first-time manager's arsenal. Most organizations require it in some form. Far fewer managers do it well. Some treat it as a formality: a box to check, a signature to collect, a 15-minute meeting that says nothing the employee didn't already know. Others skip it entirely and tell themselves they'll catch up at the annual review.

Both approaches are mistakes. The mid-year check-in, done right, is the single most powerful thing you can do to ensure the annual review is not a surprise. It is also the earliest point in the performance cycle where you can correct a trajectory before it becomes a permanent record. An employee who is underperforming in June still has six months to turn it around. An employee who finds out in December that they've been underperforming all year has been given no such opportunity.

This chapter gives you the mid-year framework, the template, the conversation guide, and the real-world language to make this check-in genuinely useful: for your employees, for your documentation, and for yourself.

# The Difference Between an Annual Review and a Mid-Year Check-In

Understanding what the mid-year is not is as important as understanding what it is. It is not a mini annual review. It is not a preliminary rating. It is not a warning document disguised as a check-in. It is a forward-facing conversation that uses the first half of the year as its data set and the second half as its opportunity.

| Annual Review | Mid-Year Check-In |
|---|---|
| Comprehensive evaluation of the full year | Focused snapshot of the first half of the year |
| Formal rating assigned and recorded | No formal rating; directional assessment only |
| Backward-looking: what happened? | Forward-looking: what needs to happen? |
| Feeds compensation, promotion, and succession decisions | Feeds the second half of the year and informs the annual review |
| Full documentation required in HR system | Brief written record, such as a simple email or shared document |
| Takes several hours to prepare and write | Takes 20–30 minutes to prepare and conduct |
| Employee often nervous or defensive | Employee typically more open, as stakes feel lower |
| Primary audience: HR, leadership, legal record | Primary audience: you and the employee, together |

The lower stakes of the mid-year check-in are actually its greatest asset. Because there is no formal rating on the line, employees are more willing to be honest. They will tell you things in a mid-year conversation they would never say in an annual review. What is frustrating them. What they feel they have not been given the resources to do. Where they feel they are genuinely excelling and not getting credit. That information is gold, and the mid-year is often the only place you will get it.

> *The mid-year check-in is the place where the annual review is won or lost. Not at the end of December when you are writing it. It is won or lost in June, when you had the conversation that either changed the trajectory or let it continue unchanged.*

# When to Do It — and When Not to Skip It

The right time for a mid-year check-in is at the midpoint of the formal review cycle. If your organization runs a January-to-December annual review, the check-in should happen in June or early July, and no later. If your cycle is April to March, the check-in falls in September or October. The exact month matters less than the principle: halfway through the year, while there is still enough time to act on what you find.

Two situations in which managers most commonly skip the mid-year check-in, and why both are mistakes:

**"Everything is going fine — I don't want to create an unnecessary meeting."**

When things are going well is precisely the right time for a mid-year check-in. Your top performers want to know you see them. They want to discuss what comes next. They want to feel that their manager is investing in their development, not just managing their output. A 25-minute conversation with a high performer mid-year costs almost nothing. Losing that high performer because they felt invisible costs an enormous amount, including in recruiting, in training, in operational continuity, and in team morale.

**"Things are not going well — I don't want to have a difficult conversation."**

This is the most damaging reason to skip a mid-year check-in, and it is also the most common. A manager who avoids a performance conversation in June because it feels uncomfortable is guaranteeing a much harder conversation in December, and removing the employee's chance to course-correct. Worse, if that employee is ultimately terminated without any mid-year documentation, the organization's legal position is significantly weaker than it would have been.

The discomfort of a mid-year performance conversation is temporary. The consequences of avoiding it are not.

> *A note from the field: In over two decades of leading distribution operations, the managers who had the fewest annual review surprises, for their employees and for themselves, were the ones who treated the mid-year check-in as non-negotiable. Not a courtesy. Not a calendar placeholder. A required conversation that they prepared for as seriously as they prepared for the annual.*

# How to Prepare for the Mid-Year Check-In

Unlike the annual review, the mid-year check-in does not require hours of preparation. But it does require intentional preparation. The difference between a meaningful conversation and a wasted 25 minutes is usually whether the manager sat down for 20 minutes the day before and actually thought about the employee.

**Your 20-Minute Prep Checklist**

- Pull the goals you set together at the start of the year. Read them. For each one, ask yourself: where are we? On track, ahead, behind, or stalled?
- Look at your notes from the year so far. If you have been documenting observations throughout the year, as this book strongly encourages, pull those notes and look for patterns. What have you seen more than once?
- Think about the person specifically. Not their role, but the person. What has been happening for them this year? Have there been changes in their circumstances, their energy, their engagement? What is your gut telling you that the data hasn't yet named?
- Decide what the one or two most important things are that you want this employee to know coming out of this conversation. Not everything, just the one or two things. The check-in that tries to cover everything usually lands nothing.
- Prepare your opening question. Not 'how's it going?' The check-in that starts with 'how's it going?' usually produces 'fine.' Start with something that requires an actual answer.

Strong opening questions for a mid-year check-in include:

- What has been the most meaningful work you've done in the first half of this year, and why?
- Where do you feel the most friction in your role right now? What is getting in your way?
- If you could change one thing about how we're working together, what would it be?
- When you think about the second half of this year, what excites you, and what concerns you?
- Looking at the goals we set in January, where do you feel confident about where we're headed, and where do you feel we need to make adjustments?

Notice what these questions have in common: they are open-ended, they require the employee to reflect rather than report, and they signal that the manager is genuinely interested in the answer. The employee who feels genuinely heard in a mid-year check-in is far more receptive to developmental feedback than one who feels they are being evaluated.

# The Mid-Year Check-In Template

The following template is designed to be completed in 20–25 minutes of actual writing time, either before or after the conversation. It is intentionally shorter than the annual review template. Its purpose is not comprehensive documentation. Its purpose is a useful, honest record of where this employee stands at the halfway point, along with what the second half of the year needs to look like.

This template can be used as a printed form, adapted into your organization's HR system, or converted into a simple shared document between manager and employee. What matters is that the conversation happened and the key points were captured.

## Section 1 — Employee and Review Basics

### Employee / Role / Department

▶ *Full name, current title, and department or team.*

### Review Period Covered

▶ *The portion of the year covered by this check-in. Example: January – June.*

### Check-In Date and Format

▶ *Date of the conversation. Note whether it was in-person, video, or phone.*

## Section 2 — Goal Status Review

For each goal set at the start of the year, assign a status and write a brief assessment. Use the four-point scale below. This is not a rating. It is a navigational tool that shows both manager and employee where they are and what direction they are heading.

| | |
|---|---|
| **On Track** | Performance against this goal meets or exceeds the pace expected at this point in the year. |
| **Needs Attention** | Progress has been made but the goal is at risk of not being met by year-end without adjustment. |
| **Off Track** | Insufficient progress has been made. Immediate course correction is required. |
| **Modified / On Hold** | This goal has been formally adjusted or paused due to changed circumstances. |

### Goal 1 — Name and Current Status

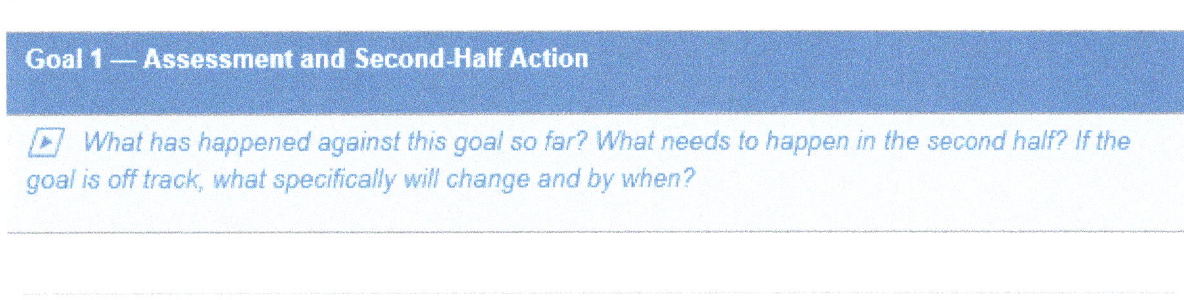


### Goal 1 — Assessment and Second-Half Action



### Goal 2 — Name and Current Status

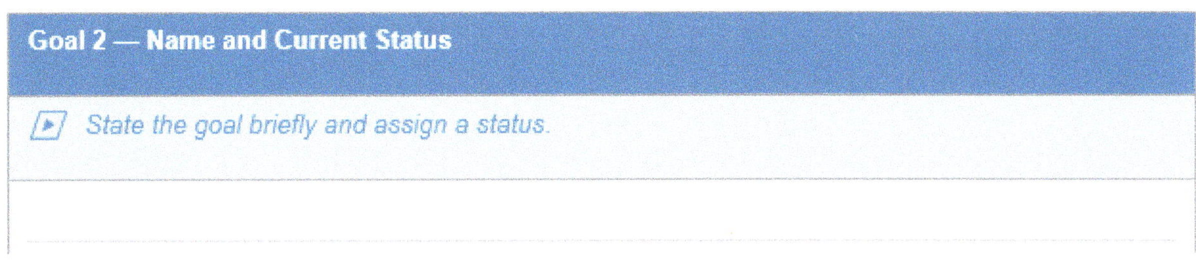

---

### Goal 2 — Assessment and Second-Half Action

▶ *What has happened and what needs to happen? Name the specific next steps.*

---

### Goal 3 — Name and Current Status

▶ *State the goal briefly and assign a status.*

---

### Goal 3 — Name and Current Status

▶ *State the goal briefly and assign a status.*

---

### Goal 3 — Assessment and Second-Half Action

▶ *What has happened and what needs to happen? Name the specific next steps.*

## Section 3 — What Is Working

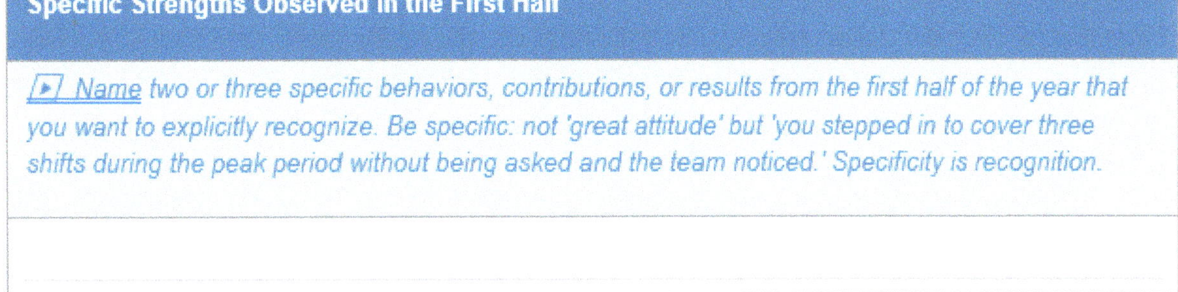

**Specific Strengths Observed in the First Half**

*Name two or three specific behaviors, contributions, or results from the first half of the year that you want to explicitly recognize. Be specific: not 'great attitude' but 'you stepped in to cover three shifts during the peak period without being asked and the team noticed.' Specificity is recognition.*

## Section 4 — What Needs to Change

This is the section most managers soften excessively. Resist the urge. An employee who receives unclear developmental feedback at mid-year has been set up to fail at year-end. Name the gap directly. Name what it looks like when the gap is closed. Name what support is available. Then let the employee own the work.

**Primary Development Area for the Second Half**

*What is the single most important thing this employee needs to work on in the next six months? Name the specific behavior or result gap. Do not list five things; prioritize the one that matters most. If you name five things with equal emphasis, the employee will work on none of them.*

**What Success Looks Like**

*How will you and the employee know that improvement has occurred? Name the observable behavior or measurable result that would indicate the gap has been closed. Vague development targets produce vague results.*

**Support Available**

*What resources, coaching, experiences, or relationships are available to help this employee improve? Be specific. 'Support is available' is not a support plan. 'We will meet every other week to review progress and I will connect you with the senior manager in the Seattle facility for a development conversation' is a support plan.*

## Section 5 — Employee Voice

This section exists because the check-in is a two-way conversation, not a broadcast. The employee's perspective on their own year, including their frustrations, their aspirations, and their observations about what is working and what isn't, is data. Record it. It demonstrates that you listened. It also protects you if the annual review is later disputed, because you have documented that the employee had the opportunity to share their perspective.

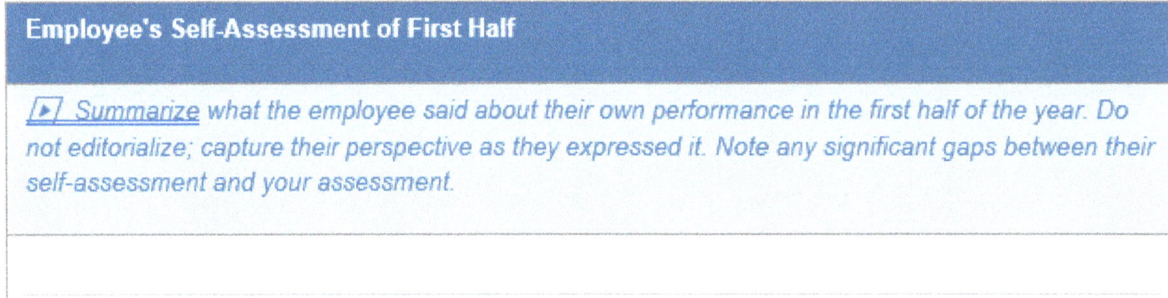


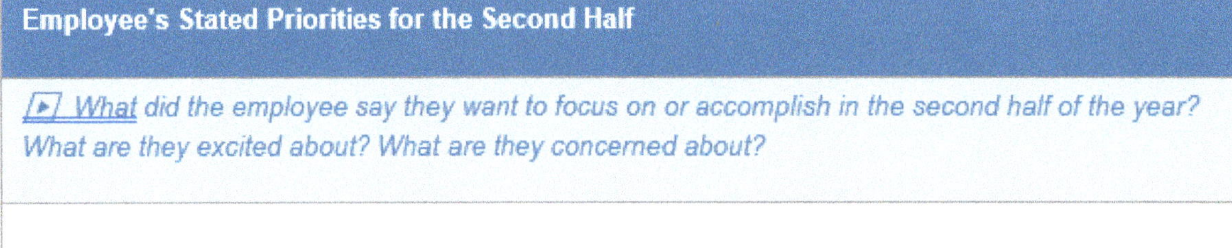


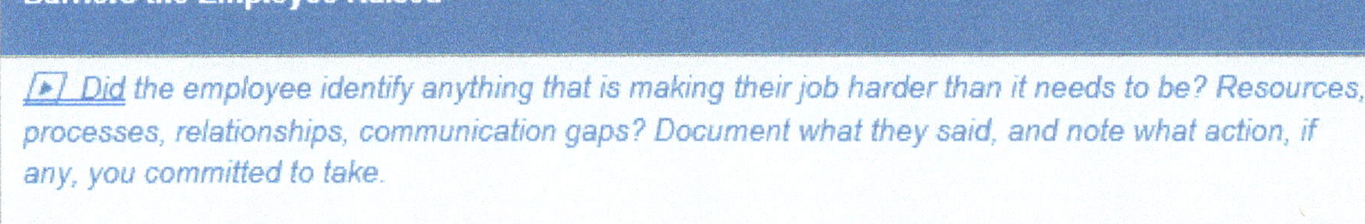

## Section 6 — Agreements and Next Steps

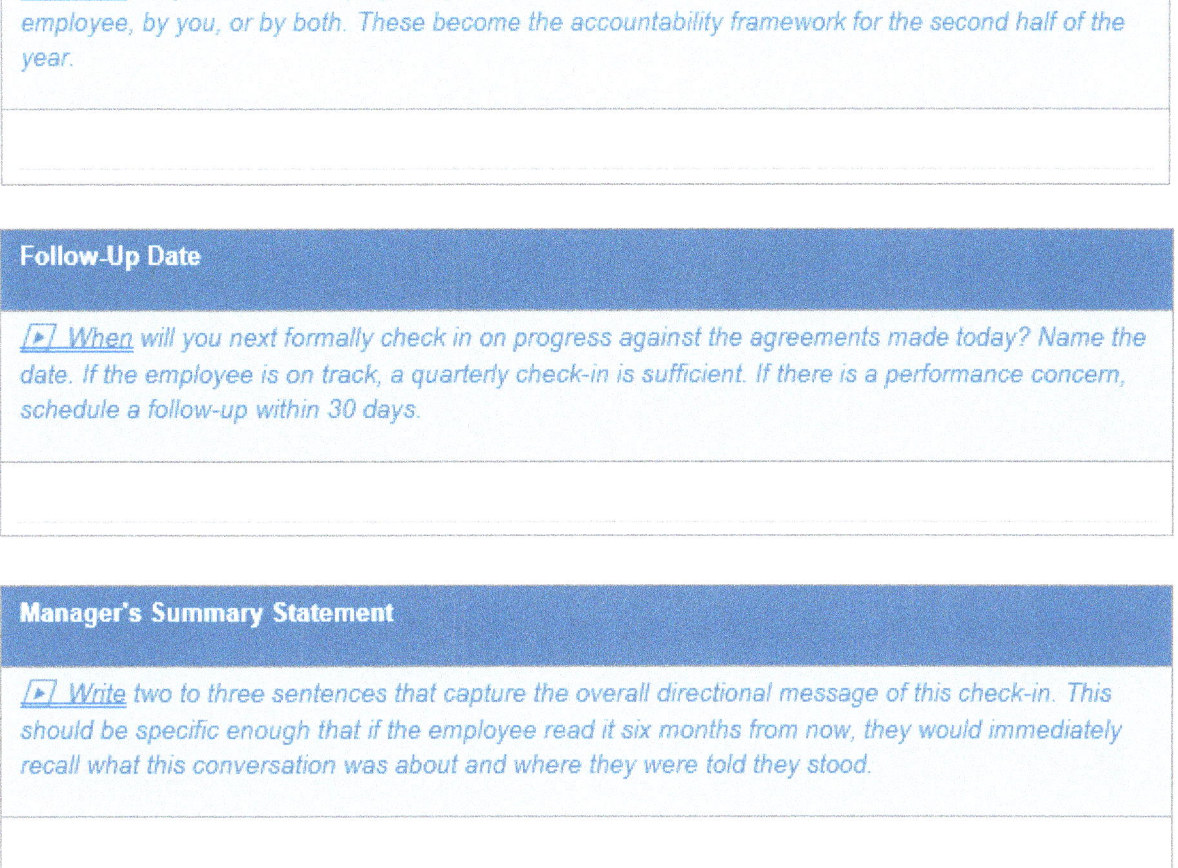

**Key Agreements from This Conversation**

*▶ What did you and the employee agree to? List the specific commitments made, whether by the employee, by you, or by both. These become the accountability framework for the second half of the year.*

**Follow-Up Date**

*▶ When will you next formally check in on progress against the agreements made today? Name the date. If the employee is on track, a quarterly check-in is sufficient. If there is a performance concern, schedule a follow-up within 30 days.*

**Manager's Summary Statement**

*▶ Write two to three sentences that capture the overall directional message of this check-in. This should be specific enough that if the employee read it six months from now, they would immediately recall what this conversation was about and where they were told they stood.*

# Sample Mid-Year Check-In Language

The following examples demonstrate how mid-year check-in language differs from annual review language in tone, scope, and forward focus. Each example is drawn from real check-in conversations across the distribution network, anonymized to protect the individuals involved.

### Example A: Employee On Track — Top Performer



First half of the year summary:

The Burlington facility is operating at a high level across every key metric. Productivity is tracking above the prior year at this point in the cycle. Same-day service is consistent. The leadership team you have built is functioning with increasing autonomy, which is exactly what we want to see at this stage of their development.

What is working:
Your ability to forecast operational obstacles before they become problems is what separates you from your peers at this level. You see the chessboard moves ahead. That skill is rare. Use it more deliberately in the second half: with your peers, not just with your team.

Development focus for the second half:
The feedback from the prior annual review stands: the leadership you demonstrate inside your four walls need to be visible outside of them. Your peers need to hear from you in group settings. That is the next step in your growth as a leader in this organization.

Agreement from this conversation:
You will take a visible role in leading one portion of the Q3 manager summit agenda. We will check in on this at the end of Q3.

**Kevin Baker**

## Example B: Employee Needs Attention — Quality Gap

**DC Manager, Texas | Goal Status: On Track (2 goals) / Needs Attention (1 goal)**

First half of the year summary:

There is real momentum in Dallas. The culture is shifting, productivity is trending upward, and the team is beginning to believe in where the building is going. That progress is visible and it belongs to you and your supervisors.

What is working:
Your communication with the peer group has been excellent. You have built trust with your counterparts in the network faster than most managers do in their first full year. That relational credibility will serve you well as we move into a more integrated network-wide approach in the second half.

What needs to change:
Quality, specifically error rates and miss-shipment frequency, remains above acceptable levels. This has been a consistent conversation. The goal is a 25% reduction in inventory adjustments by year-end. At the current rate, we are not on pace to get there. In the second half, this needs to be the primary operational focus.

What success looks like by the annual review:
Errors down 20% or more from the first-half baseline. Accountability conversations happening at the supervisor level, not just the manager level.

Agreement from this conversation:
Weekly quality review meeting added to the Dallas leadership team agenda.
Progress update at the end of Q3 with specific metrics reviewed.

## Example C: Employee Off Track — Direct Conversation Required

DC Supervisor, Illinois | Goal Status: Off Track (primary goal)

First half of the year summary:

This conversation is being held because the first half of the year has not met the
expectations of the supervisor role. That needs to be stated clearly at the start
so that what follows is understood in context.

The specific gap: attendance and schedule reliability have not met the standard
required for a supervisory position. The team has been impacted. Other supervisors
have absorbed coverage that was not their responsibility. This pattern has been
discussed twice since January and has not shown sustained improvement.

What is also true:
The associate relationships you have built are genuine and they matter. Your team
trusts you. That relational foundation is real and it is worth protecting. The way
to protect it is to show up for your team, consistently, reliably, every shift.

What the second half needs to look like:
Zero unexcused absences or late arrivals in the next 90 days. This is not a
stretch goal. It is the baseline expectation of the role. If this standard is
not met consistently by the time of the annual review, a formal performance
improvement plan will follow.

Support available:
Bi-weekly check-ins with me. Direct connection to the associate assistance
program if personal circumstances are contributing to the attendance issue.

This conversation was documented and both parties received a copy.

# The Follow-Up Email: Making the Conversation Stick

The mid-year check-in ends when you leave the room or close the video call. The follow-up email is what makes it last. It does not need to be long. It needs to be specific, timely, and honest: a brief record of what was said and what was agreed to.

Send it within 24 hours. Any longer and the conversation has already begun to fade. The email should cover three things: what you recognized, what you asked them to focus on, and what you committed to. That's it. Three elements. Most managers can write this in ten minutes.

## Sample Follow-Up Email — On Track Employee

**Sample Email | DC Manager, Washington State**

Subject: Mid-Year Check-In, Summary and Next Steps

Hi [Name],

Thank you for the conversation today. I wanted to send a quick summary of what we discussed so we have a shared record to refer back to.

Where you stand at the midpoint: strong. All three goals are on track, the team is operating well, and the customer metrics in Seattle continue to be among the best in the network. That reflects your leadership and your team's dedication to the standard you have set.

What we agreed to focus on in the second half: your visibility and engagement in group settings with your peers. You know what you know. The organization benefits when you share it more broadly. We talked about taking a visible role in the Q3 summit agenda, and I am expecting that to happen.

My commitment to you: a dedicated 30 minutes at the end of each quarter to discuss your development and what you need from me.

Good first half. Let's finish well.

[Your name]

## Sample Follow-Up Email — Performance Concern



Subject: Mid-Year Check-In, Summary and Expectations for the Second Half

Hi [Name],

Per our conversation today, I am summarizing the key points in writing so that we have a shared and documented record of what was discussed.

The primary concern discussed: attendance and schedule reliability have not met the standard expected of a supervisor during the first half of this year. This was discussed on [prior dates] and today represents the third formal conversation on this topic.

The expectation for the second half is clear: zero unexcused late arrivals or absences in the next 90 days. This is the baseline standard of the role, not an elevated expectation. Consistent performance at this level by the time of the annual review is required. Failure to demonstrate sustained improvement will result in a formal performance improvement plan.

What is also true and what I said directly today: your associate relationships are a genuine strength. I want to see you succeed in this role. The path to doing that runs through reliability, which means showing up for your team the same way they show up for you.

Our next check-in is scheduled for [date, 30 days from now].

If you have any questions about what was discussed or anything in this email, please reach out.

[Your name]

## Chapter 10 — Key Takeaways

• The mid-year check-in is not a mini annual review. It is a forward-facing conversation that uses the first half of the year as its data.

• The best time to do a mid-year check-in is when you think you don't need one. Top performers especially deserve this investment.

• 20 minutes of intentional preparation the day before produces a dramatically better conversation than walking in cold.

• Start with a question that requires a real answer, not 'how's it going?' but 'where do you feel the most friction right now?'

• The development section is the hardest to write honestly. Write it honestly anyway. Vague mid-year feedback produces vague second-half results.

• Send the follow-up email within 24 hours. It makes the conversation official, demonstrates that you listened, and creates the paper trail that protects everyone.

• The mid-year check-in is where the annual review is won or lost. Treat it accordingly.

# Chapter 11

# The Goal-Setting Template for the Year Ahead

*A goal without a measure is a wish. A measure without context is a number. A goal that connects what an individual does every day to what the organization is trying to become is something worth working toward. Setting that kind of goal is harder than it looks. It is also the most important thing you will do for your employees at the start of every year.*

The performance cycle does not begin with the review. It begins with the goals that are set before the year starts. Every performance review is, at its core, an assessment of how well the employee performed against what was agreed upon twelve months earlier. If the goals were vague, unmeasurable, or disconnected from what the employee actually does, the review will be equally vague, no matter how well you write it.

This chapter is the beginning of the loop. Earlier chapters address how to document what happens during the year and how to write the performance assessment at the end. This chapter lays the foundation that makes both of those possible. Good goals are the architecture of a good year. They tell the employee what success looks like. They give the manager a clear standard to evaluate against. And they connect individual effort to the larger purpose of the team and the organization.

This chapter covers what makes a goal strong versus weak, how to build buy-in from the employee so the goal feels owned rather than imposed, how to connect individual goals to team and organizational objectives, and a complete goal-setting template with real-world examples drawn from distribution operations at every level, from hourly supervisor to senior manager.

## Why Most Goals Fail Before the Year Begins

The majority of performance goals fail not because the employee could not achieve them, but because the goals were never designed to be achievable in any meaningful sense. They were written quickly, at the end of a review meeting when everyone was tired, to satisfy a form requirement. They used generic language. They had no measurement criteria. They were handed to the employee rather than built with them. And they were filed away and not looked at again until December.

Here are the five most common goal-setting failures, and what they cost.

## Failure 1: The Goal That Cannot Be Measured

"Improve communication skills." "Become a better leader." "Focus on quality." These are intentions, not goals. A goal that cannot be measured cannot be fairly evaluated. When December comes and you sit down to rate this employee on whether they 'improved communication skills,' you will be rating an impression, not a result. And impressions are not documentation.

## Failure 2: The Goal That Is Entirely Outside the Employee's Control

"Increase facility revenue by 15 percent." If the employee has no direct influence over pricing, customer acquisition, or sales strategy, this goal is not a performance target. It is a weather forecast. Goals must be tied to things the employee can actually influence through their daily decisions, behaviors, and leadership. Tying someone's rating to factors outside their control breeds resentment and destroys the credibility of the entire review process.

## Failure 3: The Goal That Was Never Revisited

A goal set in January and never discussed again until December is not a goal. It is an artifact. Goals require ongoing attention. They should be referenced in one-on-ones, reviewed at the mid-year check-in, adjusted if circumstances change, and tracked consistently throughout the year using a reliable documentation system. A goal that disappears into a file drawer the day after it is written serves no one.

## Failure 4: The Goal That Was Imposed Rather Than Built

When an employee has no input into their own goals, they do not own them. They comply with them, or they don't. The research on goal commitment is unambiguous: people work harder toward goals they helped create. This does not mean employees get to set any goals they want. It means the conversation about what the goal will be is a real conversation, not a performance. The manager brings the organizational expectation. The employee brings the operational reality. The goal emerges from the intersection.

## Failure 5: Too Many Goals

Three to five well-written goals will drive performance. Eight to twelve goals will paralyze it. When everything is a priority, nothing is. The first-time manager who tries to capture every responsibility in a goal list has not written goals. They have written a job description. Goals should represent the highest-leverage things the employee can do to grow, improve, and contribute at the next level. Not everything they do. The most important things they can do.

> *The goal-setting conversation and the review conversation are bookends of the same story. If you write the goals with care in January, the review in December will write itself. If you write the goals carelessly in January, no amount of skill in December will compensate.*

# The SMART-Plus Framework — What Every Strong Goal Requires

The SMART framework, which stands for Specific, Measurable, Achievable, Relevant, and Time-Bound, has been the standard goal-writing model in management for decades, and it remains useful because its elements are genuinely necessary. A goal that is missing any one of the five elements is a weaker goal. But in operational leadership, two additional elements matter enough to earn their own place in the framework.

| Letter | Element | The Question It Answers | Example (Operations Context) |
|---|---|---|---|
| S | Specific | What exactly will be accomplished? Who is responsible? What actions will be taken? | *Reduce inbound receiving errors in the pick zone by improving associate training and process adherence.* |
| M | Measurable | How will we know it has been achieved? What metric will we track and what is the target? | *Reduce inventory adjustment dollars by 20% from the prior year baseline of $800,000.* |
| A | Achievable | Is this realistic given current resources, staffing, and operational conditions? Has it been done before? | *Prior year reduction of 12% was achieved with similar resources. 20% is a stretch but attainable.* |
| R | Relevant | Does this goal connect to team and organizational priorities? Does it matter for this employee's role? | *Inventory accuracy directly impacts customer satisfaction scores and operational cost targets.* |
| T | Time-Bound | When will this be achieved? Are there interim milestones to track progress during the year? | *10% reduction by mid-year check-in (June). Full 20% reduction by December 31.* |
| + | Owned | Has the employee participated in setting this goal? Do they understand why it exists and believe it is fair? | *Goal discussed in January goal-setting meeting. Employee agreed to the target and identified two process changes they will drive.* |
| + | Connected | Does this goal link explicitly to a team or organizational objective above the individual level? | *Supports the network-wide Perfect Pallet Initiative and the company's customer retention strategy.* |

The Owned and Connected elements are what separate goals that employees work toward from goals that employees work around. When an employee understands why their goal matters to the organization, not just what they are supposed to do, the goal becomes part of how they think about their work rather than a compliance requirement they check off. That shift in ownership is the difference between a goal that drives performance and a goal that generates paperwork.

**Kevin Baker**

# Weak Goals vs. Strong Goals — Side by Side

The following comparisons are drawn from real goal-setting language across distribution operations at multiple levels. Each pair shows the same intent written two ways: one that will fail to drive performance, and one that will not.

## Productivity and Operations

| ✗ Weak Goal — Not Measurable | ✔ Strong Goal — Specific and Measurable |
| --- | --- |
| *Improve productivity in the facility.* | *Increase facility cartons processed per hour worked (CPHW) from the current baseline of 38 to a sustained average of 42 by Q4, measured weekly. Interim target: 40 CPHW by the mid-year check-in.* |

## Quality and Accuracy

| ✗ Weak Goal — Not Measurable | ✔ Strong Goal — Specific and Measurable |
| --- | --- |
| *Reduce errors in the pick zone.* | *Reduce inventory adjustment dollars in the inbound processing area by 25% from the prior year baseline of $800,000. Track monthly. Identify root causes of the top three error types by end of Q1 and implement process changes by Q2.* |

## Team Development and Succession

| ✗ Weak Goal — Not Measurable | ✔ Strong Goal — Specific and Measurable |
| --- | --- |
| *Develop your supervisors.* | *Complete Individual Development Plans (IDPs) for all three supervisors by February 15. Each IDP must include at least two actionable development activities with specific timelines. Conduct a formal development check-in with each supervisor quarterly and document the conversation.* |

## Self Development

| ✘ Weak Goal — Not Measurable | ✔ Strong Goal — Specific and Measurable |
|---|---|
| *Work on your professional development.* | *Complete one relevant professional development course through the Learn On platform by Q3. Present key learnings to the peer group at the Q3 manager summit. Additionally, complete two customer or sales ride- along   by year-end and submit a written reflection of observations within one week of each.* |

## Safety

| ✘ Weak Goal — Not Measurable | ✔ Strong Goal — Specific and Measurable |
|---|---|
| *Maintain a safe workplace.* | *Achieve zero OSHA recordable incidents for the full year. Conduct weekly safety walks with documented findings and same-day corrective action for any violations found. Maintain the safety observation log and review with the team monthly. Target: Chasing Zero gold-level recognition by Q4.* |

## Customer and Service

| ✘ Weak Goal — Not Measurable | ✔ Strong Goal — Specific and Measurable |
|---|---|
| *Improve customer service metrics.* | *Maintain same-day shipped percentage at or above 98.5% for the full year, measured weekly. When service falls below 97%, a root cause analysis must be completed within 48 hours and shared with the manager. Target the top three customer service complaint categories from last year for specific process improvements by Q2.* |

# Connecting Individual Goals to the Bigger Picture

One of the most powerful things a manager can do when setting goals is to show the employee explicitly how their individual target connects to the team's goals, and how the team's goals connect to the organization's strategy. This is not corporate theater. It is genuine motivational architecture.

People work harder when they understand why their work matters beyond their immediate responsibility. An associate who knows that reducing pick errors contributes to customer retention, which contributes to the organization's ability to stay competitive, has a reason to care about accuracy that goes beyond avoiding a correction. That reason is what transforms a goal from a number on a form into a purpose.

## The Three-Level Connection

When writing or discussing any goal, be able to articulate all three levels:

| | |
|---|---|
| **Level 1: Individual** | What this specific employee will do and achieve. |
| **Level 2: Team** | How this goal contributes to the team's collective objectives and metrics. |
| **Level 3: Organization** | How the team's performance connects to the company's strategic priorities: customer satisfaction, cost efficiency, market position, culture. |

Here is what that three-level connection looks like in practice for a single goal:

**Three-Level Goal Connection — Quality / Inventory Accuracy**

### Goal Statement

*Reduce inventory adjustment dollars in the inbound processing area by 25% from the prior year baseline of $800,000. Track monthly against a $600,000 year-end target.*

### How We Measure It

*Monthly inventory adjustment report, tracked against the $800,000 baseline. Mid-year target is $700,000 or below. Results reviewed at every monthly operations meeting and shared with the full team.*

### How It Connects Up

*This goal feeds directly into the DC Manager network's collective accuracy target, which rolls up to the Perfect Pallet Initiative, the organization's primary quality strategy for the year. Every dollar of inventory adjustment is a dollar that comes out of operational margin and potentially out of customer confidence.*

### Why the Employee Owns It

*This employee leads the team that processes inbound shipments. The quality of their team's work, including the training they deliver, the processes they enforce, and the standards they model, is the primary driver of this result. No one else has more control over this number.*

When you can explain a goal at all three levels, and when you explain it that way to the employee during the goal-setting conversation, the goal stops being something imposed on them and becomes something they can see themselves in. That visibility is the foundation of ownership.

## The Goal-Setting Conversation — How to Build Buy-In

The goal-setting template in the next section is a document. But the document is only as useful as the conversation that produces it. A goal-setting meeting where the manager reads goals off a list and asks the employee to sign them is not a goal-setting meeting. It is a compliance exercise. The conversation that actually builds commitment looks different.

**Before the Meeting**

Send the employee the organizational and team priorities at least three days before the goal-setting meeting. Ask them to come prepared with two questions: What do they think their highest-leverage contribution will be this year? And what obstacles do they anticipate?

## During the Meeting

Start by asking, not telling. What do you see as the most important things you can work on this year? What areas did you feel the prior year's goals missed or misrepresented your actual responsibilities? Where do you feel there is the most opportunity for you to grow? Listen to the answers before you share your own view. In most cases, a good employee will identify the same priorities you did. When they name it themselves, they own it in a way they cannot when you name it for them.

## The Negotiation

There will be goals where you and the employee disagree on the target. That is normal. The manager has organizational context the employee may lack. The employee has operational context the manager may lack. A target of 52 cartons per hour may sound achievable from the organizational level and unrealistic given a specific staffing constraint the employee knows about. That tension is productive. Work through it. Do not simply override the employee's concern. Do not simply capitulate to it. Find the standard that is genuinely ambitious and genuinely fair.

## Closing the Conversation

End every goal-setting meeting by asking the employee to restate the goals in their own words. Not read them back, but restate them. What they say tells you whether they understood and whether they are genuinely bought in. A goal they can explain in their own words is a goal they own. A goal they cannot explain is a goal that will not drive performance.

> *The best goal-setting conversations feel less like a manager assigning targets and more like two professionals agreeing on what success looks like. When you get it right, the employee leaves the room with something to work toward. When you get it wrong, they leave with something to comply with. The difference in performance between those two outcomes is not small.*

## The Goal-Setting Template

The following template is designed to be completed collaboratively, by manager and employee together, during or immediately following the goal-setting conversation. One copy for the employee. One copy for your documentation file. Both signed.

## Section 1 — Employee and Year Information

**Employee Name / Role / Department**

▶ *Full name, current title, and team or department.*

**Review Year**

▶ *The full calendar or fiscal year these goals cover.*

**Goal-Setting Meeting Date**

▶ *The date the goal-setting conversation took place.*

**Organizational Priorities This Year**

▶ *List the two or three highest-level organizational or team priorities that these individual goals should connect to. This keeps the individual goals anchored to the bigger picture throughout the year.*

## Section 2 — Individual Goals

Complete one full block for each goal. Three to five goals is the recommended range. More than five goals dilutes focus. Fewer than three may not adequately cover the employee's full scope of responsibility.

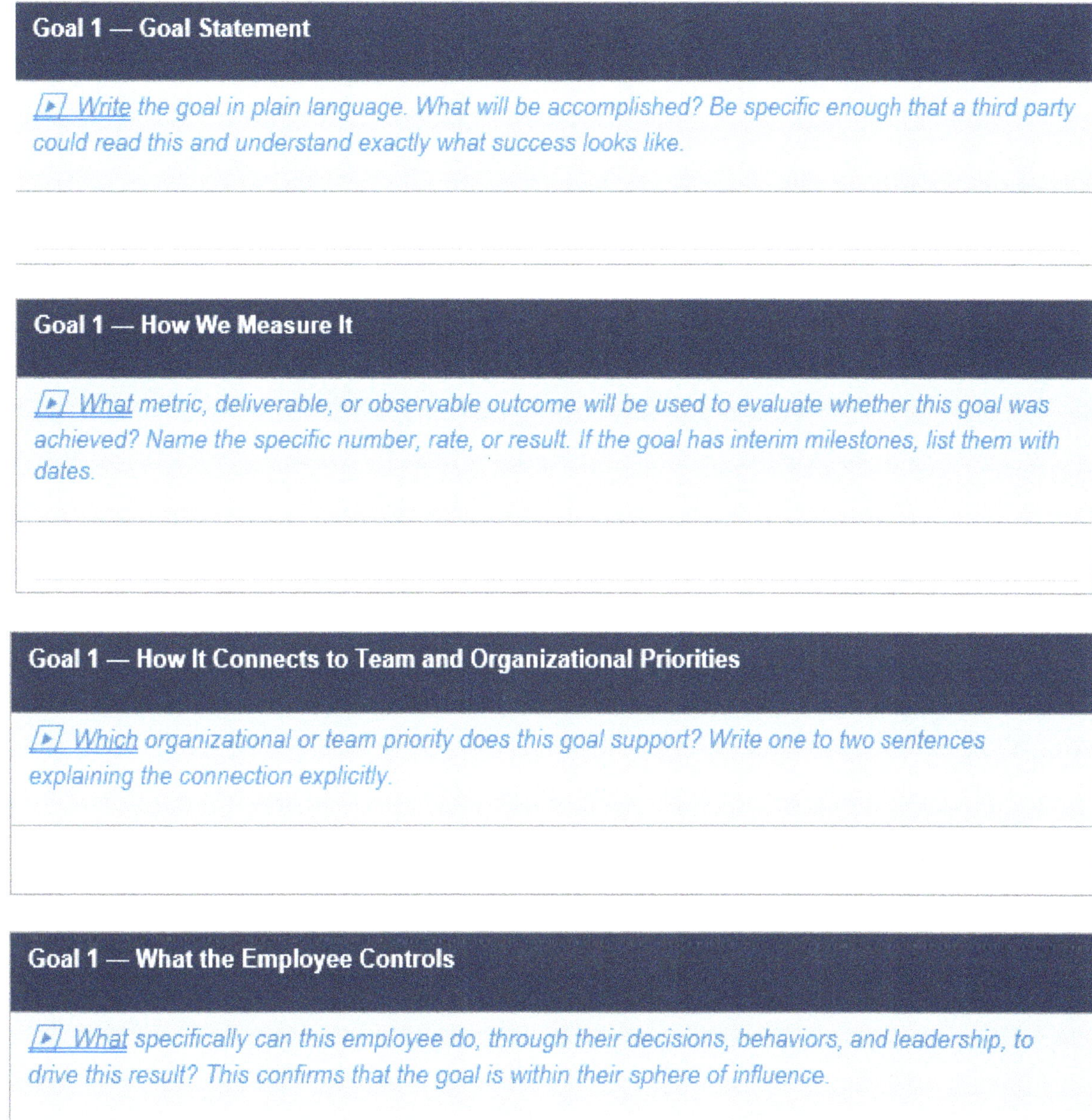

**Goal 1 — Goal Statement**

▶ *Write the goal in plain language. What will be accomplished? Be specific enough that a third party could read this and understand exactly what success looks like.*

**Goal 1 — How We Measure It**

▶ *What metric, deliverable, or observable outcome will be used to evaluate whether this goal was achieved? Name the specific number, rate, or result. If the goal has interim milestones, list them with dates.*

**Goal 1 — How It Connects to Team and Organizational Priorities**

▶ *Which organizational or team priority does this goal support? Write one to two sentences explaining the connection explicitly.*

**Goal 1 — What the Employee Controls**

▶ *What specifically can this employee do, through their decisions, behaviors, and leadership, to drive this result? This confirms that the goal is within their sphere of influence.*

**Goal 1 — Obstacles and Resources Needed**

▶ *What barriers might prevent this goal from being achieved? What resources, support, or access does the employee need? Document commitments made by the manager here.*

**Goal 2 — Goal Statement**

▶ *Write the goal in plain language. Specific, measurable, achievable, relevant, time-bound.*

**Goal 2 — How We Measure It**

▶ *What metric or deliverable defines success? Include interim milestones if applicable.*

**Goal 2 — How It Connects Up**

▶ *Which organizational or team priority does this goal support?*

**Goal 2 — What the Employee Controls**

▶ *What can this employee specifically do to drive this result?*

## Goal 2 — Obstacles and Resources Needed

▶ *What might get in the way? What support is the manager committing to provide?*

## Goal 3 — Goal Statement

▶ *Write the goal in plain language.*

## Goal 3 — How We Measure It

▶ *What metric or deliverable defines success?*

## Goal 3 — How It Connects Up

▶ *Which organizational priority does this goal support?*

## Goal 3 — What the Employee Controls

▶ *What actions or decisions by this employee drive this result?*

## Goal 3 — Obstacles and Resources Needed

▶ *Barriers, support needed, and manager commitments.*

If additional goals are needed, repeat the five-field block above for Goal 4 and Goal 5.

## Section 3 — Development Goal

Every employee, regardless of performance level, should have at least one development goal. This is separate from performance goals. It is focused on the employee's professional growth: a skill to build, a capability to develop, or an experience to pursue. Development goals are not about fixing weaknesses. They are about building the next version of this employee.

**Development Focus Area**

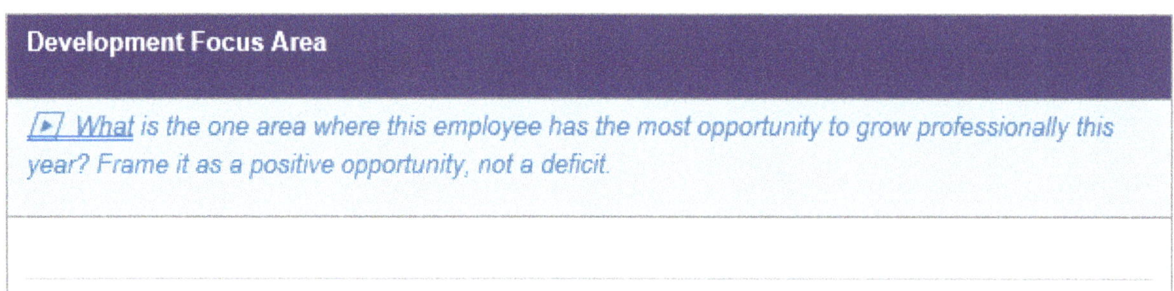


**Development Activities**

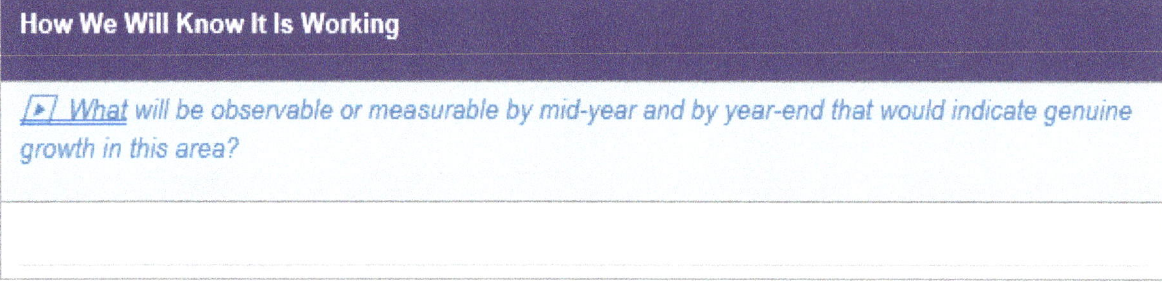


**How We Will Know It Is Working**

What will be observable or measurable by mid-year and by year-end that would indicate genuine growth in this area?

## Section 4 — Agreements and Sign-Off

**Employee Commitment Statement**

*▶ In the employee's own words: what do these goals represent and what are they committing to? This can be written by the employee or captured verbatim from the goal-setting conversation.*

**Manager's Commitment Statement**

*▶ What is the manager committing to provide in support of these goals? Resources, coaching, access, information, advocacy. Write it down. If you commit to it verbally in the goal-setting meeting, it belongs in this field.*

**First Check-In Date**

*▶ When will the manager and employee next formally review progress against these goals? This should be no more than 90 days from the goal-setting meeting. Write the date here and put it on both calendars before the meeting ends.*

# Real-World Goal Examples Across Role Levels

The following examples are drawn from actual goal documents used across a multi-facility distribution network. They are anonymized and organized by role level to show how goals should differ in scope and complexity as responsibility increases.

## Hourly Supervisor Level

### Goal: Team Productivity | DC Supervisor, Illinois

**Goal Statement**

Lead the outbound shipping team to achieve and maintain an average of 48 cartons per hour worked across all shifts by Q3, with a stretch target of 52 by year-end.

**How We Measure It**

Weekly CPHW report reviewed every Monday morning. Supervisor receives individual team metric each week. Results discussed at biweekly one-on-one with the DC Manager.

**How It Connects Up**

Team productivity is the primary driver of facility cost-per-unit performance, which feeds directly into the network's operational efficiency target for the year.

**Why the Employee Owns It**

This supervisor directs the daily work of the outbound team. The pacing decisions, staffing assignments, and in-shift coaching they provide are the primary variables that determine this result.

## DC Manager Level

### Goal: Strategic Initiative Support | DC Manager, New Jersey

**Goal Statement**

Lead the Burlington facility's implementation of the new WMS voice-picking technology by Q2. This includes associate training, process documentation, parallel testing, and go-live readiness sign-off. Achieve a post-go-live error rate no higher than the pre-implementation baseline within 60 days of launch.

**How We Measure It**

Go-live date on or before June 30. Post-go-live error rate benchmark achieved by August 31. Training completion rate for all associates: 100% by May 15. No more than two implementation-related service failures in the 30 days following launch.

**How It Connects Up**

WMS implementation is one of the organization's top three strategic priorities for the year. Successful execution in Burlington sets the standard and provides the playbook for two additional facilities implementing the same technology in Q3 and Q4.

**Why the Employee Owns It**

The DC Manager controls the training timeline, the change management approach, the associate communication, and the operational decisions around parallel testing. No one else has more influence over whether this implementation succeeds in Burlington.

**Senior Manager / Director Level**

<table>
<tr><td>Goal: Succession and Bench Development | Senior Director of Distribution</td></tr>
</table>

**Goal Statement**

*By year-end, ensure that each of the five DC Manager roles has an identified internal successor who is assessed as ready-in-12-months or better. Conduct a formal succession calibration with HR in Q2 and again in Q4. Present the succession bench to senior leadership at the Q4 review.*

**How We Measure It**

*Succession calibration completed twice per year (Q2 and Q4). Each of five facilities has a documented successor assessed at ready-in-12-months or above by December 31. At least two high-potential supervisors in the network begin a formal stretch assignment by Q3.*

**How It Connects Up**

*Bench strength is a direct risk management priority for the organization. Losing a DC Manager without a ready successor creates operational disruption, increased recruiting costs, and cultural continuity risk. A strong bench is also a retention tool, as high-potential employees who see a clear path forward are far more likely to stay.*

**Why the Employee Owns It**

*The Director of Distribution controls the coaching conversations, development assignments, calibration process, and advocacy for high-potential associates across the network. HR is a partner in the process, but ownership lives here.*

# When Goals Need to Change Mid-Year

Not every goal set in January survives contact with the actual year. Businesses change. Priorities shift. Staffing fluctuates. A goal that made complete sense in January may be irrelevant, unachievable, or entirely superseded by events by June. The question is not whether goals can be modified. They can and sometimes should be. The question is how.

## Legitimate Reasons to Modify a Goal

- A significant change in organizational strategy that makes the original goal irrelevant or counterproductive
- An operational event outside the employee's control that makes the original target impossible, such as a system failure, a major staffing loss, or a facility move
- A role change or significant expansion of scope that makes the original goals incomplete
- New information that reveals the original metric was flawed or unmeasurable in the way it was written

## Not Legitimate Reasons to Modify a Goal

- The goal turned out to be harder than expected and the employee is falling behind
- The employee prefers different goals than the ones that were agreed upon
- The manager wants to avoid a difficult conversation about underperformance
- The year has been 'tough' in a general sense that affected everyone equally

When a legitimate goal modification is warranted, document it in writing: the original goal, the reason for the modification, and the revised goal or target. Both manager and employee should acknowledge the change. This documentation protects the integrity of the year-end review by making clear why the goal changed and ensuring the evaluation is against the modified standard, not the original one.

When pressure to modify a goal is not legitimate, meaning it is driven by underperformance rather than changed circumstances, the right response is a coaching conversation, not a goal modification. Hold the standard. Provide the support. Document the gap. That is the performance management process working as it should. Modifying the goal to avoid the conversation is the process failing.

### Chapter 11 — Key Takeaways

- The review is only as strong as the goals that preceded it. Write goals carefully in January and the review in December will write itself.
- The five most common goal failures: unmeasurable, outside the employee's control, never revisited, imposed rather than built, and too many.
- Use the SMART-Plus framework: Specific, Measurable, Achievable, Relevant, Time-Bound, plus Owned and Connected.
- Every strong goal answer three questions without ambiguity: what will be achieved, how we will know when it is achieved, and when it needs to happen.
- Connect every individual goal to the team and organizational level. People work harder when they understand why their work matters beyond their immediate task.
- The goal-setting conversation is as important as the goal document. Ask before you tell. Listen before you assign. End by asking the employee to restate the goals in their own words.
- Three to five goals is the right range. More than five dilutes focus. Every employee should also have at least one development goal.
- Goals can be modified mid-year for legitimate reasons. Falling behind is not a legitimate reason. Document any modification in writing.

# Chapter 12

# How to Prepare for the Review Meeting

*Most managers spend weeks writing the review and twenty minutes preparing for the conversation. That is exactly backward. The written review is the record. The conversation is the event. How you show up in that room, your preparation, your tone, your command of the material, and your ability to listen, determines whether the review lands or whether it is forgotten by Friday.*

You have written the review. You have run the self-audit, confirmed that the language is specific, the rating matches the narrative, and nothing in the document would embarrass you if HR read it tomorrow. Now comes the part most first-time managers underestimate: sitting across from a person and delivering what you wrote with clarity, confidence, and humanity.

The performance review conversation is one of the most significant interactions a manager has with an employee all year. For many employees, it is the most important professional conversation they will have in any given twelve months. It shapes how they feel about their work, their manager, and their future with the organization. A review that is written brilliantly and delivered carelessly is not a good review. A review that is written adequately and delivered with genuine skill and care can be transformative.

This chapter covers everything that happens before you walk into the room: what to send the employee in advance, how to read your own review before you present it, how to set the physical and psychological environment, the one thing most first-time managers forget, and a complete preparation checklist you can use before every review conversation.

## The Week Before — What to Do Before You Schedule Anything

Preparation for the review meeting does not begin the morning of the meeting. It begins a week earlier with three specific actions that most first-time managers skip, and whose absence is usually visible in the conversation.

### Action 1: Read Your Own Review as If You Are the Employee

Before you do anything else, print or pull up the review you have written and read it once through from the employee's perspective. Not as the author, but as the recipient. Ask yourself: if I received this document about myself, what would I feel? What would surprise me? What would feel unfair? What would I want to push back on?

This exercise surfaces two things. First, any language that is harsher than you intended when you will hear it through the employee's ears rather than your own. Second, the questions the employee is likely to ask, which you should be prepared to answer before you sit down across from them, not for the first time in the middle of the meeting.

### Action 2: Anticipate the Three Hardest Moments

Every review has predictable friction points. Before the meeting, identify the three moments in your specific review where the conversation is most likely to get difficult. The rating the employee won't agree with. The developmental feedback they might resist. The goal they feel was unfair to begin with. Write down what you will say at each of those moments. Not a script, but a prepared thought. The manager who has thought through the hard moments in advance handles them with composure. The manager who hasn't handles them with whatever comes out under pressure.

### Action 3: Know Your One Thing

If the employee walks out of this meeting and remembers only one thing you said, what do you want it to be? Not the rating. Not the bullet point list of accomplishments. The one message that matters most for this specific person at this specific moment in their career. Name it before you walk in. Then make sure you say it clearly, directly, and without burying it in everything else you need to cover.

> *Knowing your "one thing" before you walk in is what makes the conversation one the employee remembers years later. Most managers have no idea what their one thing is, because they have not thought about it. That shows, not in what they say, but in what fails to land.*

## What to Send the Employee Before the Meeting

The question of what to share with the employee before the review meeting is one managers handle differently, and organizational policy varies. But the principle is consistent: the more informed the employee is before they walk into the room, the more productive the conversation will be.

An employee who receives the review document cold, reading it for the first time while sitting across from their manager, is processing it emotionally in real time. They cannot be fully present for the conversation because part of their brain is still reading and reacting to what they just learned. An employee who has had 24 to 48 hours with the document walks into the meeting ready to talk, not to read.

## What to Send — The Recommended Approach

- Send the completed review document 24 to 48 hours before the meeting. Include a brief note that acknowledges they will have questions and that the meeting is the place to discuss them.
- Send the self-review form at the same time, if your organization uses one. Ask the employee to complete it before the meeting and bring it to the conversation.
- Include the agenda for the meeting, keeping it simple rather than formal, so the employee knows how the time will be structured and what to expect.
- If there is significant developmental feedback or a rating the employee is likely to find surprising, consider a brief verbal heads-up before the document is sent. Not the full conversation, but just enough so the document is not a shock.

## What Not to Send

- Do not send the review and ask the employee to sign it digitally before the meeting. The signature should happen after the conversation, not before it.
- Do not send a version of the review and then change it significantly before the meeting. What they read should match what you discuss.
- Do not send it with so much advance notice that the employee has days to build anxiety, resentment, or a legal strategy before you have had the conversation.

---

**Sample Pre-Meeting Email to Employee**

*Subject: Performance Review, [Employee Name], [Date]*

*Hi [Name],*

*Your annual performance review is scheduled for [date] at [time] in [location/format]. I have attached the completed review document for you to read before our meeting.*

*Please take time to review it before we sit down together. I expect it will raise questions and I want us to have a real conversation, not a reading session. Come prepared to discuss what resonated, what surprised you, and anything you want to push back on. All of that is welcome.*

*I have also attached the self-review form. Please complete it and bring it to the meeting. Your perspective on your own year matters and it will be part of our discussion.*

*The meeting will take approximately 45 minutes. We will cover the review, your self-assessment, goals for the coming year, and your development.*

*See you then.*

*[Your name]*

---

*Notice what the email above does not say: 'Let me know if you have any questions.' That phrase deflects the conversation to email when it belongs in the room. It also signals that questions might be un5/26/2026welcome. Instead, explicitly invite the employee to bring their questions, their disagreements, and their own perspective to the meeting. That invitation changes the nature of the conversation before it begins.*

## Setting the Room — The Environment Matters More Than You Think

The physical and psychological environment of a performance review conversation is not incidental. It is part of the message. Where you sit, what is on the table between you, whether the door is closed, whether your phone is visible: all of these things signal to the employee how important this conversation is and how much you are present for it.

### Physical Setup

| ✔ Do This | ✘ Not This |
| --- | --- |
| Use a private room with a closed door, no exceptions | Conduct reviews in open areas, common spaces, or where others can hear |
| Sit at a round table or side by side if possible, not across a power-desk barrier | Sit behind your desk with the employee on the other side like a job interview |
| Have water available for both of you | Keep your laptop open and emails visible during the conversation |
| Put your phone face down or out of sight before they walk in | Check your phone during the meeting for any reason |
| Have a printed copy of the review for both of you | Read the review aloud to the employee word for word |
| Allow enough time: 45 to 60 minutes minimum, not 20 | Schedule reviews back to back with no buffer time between them |
| Hold all other notifications until the meeting is over | Start the meeting late and rush through it |

The round table or side-by-side arrangement is worth specific attention. The traditional across-the-desk configuration places the manager in a position of authority and the employee in a position of being evaluated, which is accurate, but it is not the only dynamic you want in the room. You also want a collaborative dynamic. The manager and employee are, in this moment, both looking at the same document and discussing the same year. The physical arrangement should reflect that.

**Psychological Setup: What You Bring Into the Room**

The physical environment is the easier half. The harder half is what you bring in psychologically. The manager who walks into a review meeting frustrated, distracted, or emotionally unresolved about the employee is not ready to have a productive conversation. The employee will feel that within the first two minutes, and it will shape everything that follows.

Before you walk in, take two minutes to do the following:

- Remind yourself of one genuine thing you respect or appreciate about this employee. Even if the review is difficult, there is something. Find it before you go in.
- Release whatever frustration or impatience you are carrying from the rest of your day. The review meeting deserves your full attention. What happens in the next 45 minutes may shape this person's trajectory. Treat it accordingly.
- Remind yourself of your 'one thing.' The most important message you want this person to leave with. Say it out loud to yourself if it helps.
- Commit to listening. The meeting will not go as planned. The employee will say something unexpected. Your job is not to defend the document; it is to have an honest conversation. Listening is how you do that.

# The One Thing Most First-Time Managers Forget

Ask the employee to do a self-review first.

This is the most consistently overlooked step in the review meeting preparation process, and its absence changes the entire dynamic of the conversation. When the manager opens the meeting by presenting the review they wrote, by going first, several things happen simultaneously, none of them ideal.

The employee shifts immediately into reception mode. They are listening and reacting to what the manager has said rather than bringing their own perspective to the table first. Whatever they might have said about themselves, the accomplishment they are most proud of, the challenge they found most difficult, the goal they felt was set unfairly, is now filtered through the frame of the manager's assessment. The conversation becomes a reaction rather than a dialogue.

When the employee goes first, when the manager opens the meeting by asking "Tell me about your year" or "Walk me through your self-review," something different happens. The employee owns the opening. They bring their own narrative. And the manager learns something they may not have known: how the employee sees themselves. That information is valuable whether it confirms the manager's view or reveals a significant gap.

**Kevin Baker**

## How to Open the Meeting

**Opening the Review Conversation — Manager Script**

**MANAGER: Thanks for coming in. Before we get into the review I've written,** *I want to hear from you first. You had a full year of work that I observed from the outside. You lived it from the inside. <u>So</u> let's start there.*

*Walk me through your year. What are you most proud of? Where did you feel you fell short? And if you had to name the one thing you most want me to understand about what this year was like for you, what would it be?*

**EMPLOYEE: [Employee speaks; manager listens without interrupting]**

**MANAGER: [After the employee finishes] Thank you. What you said about [specific** *point] is something I want to come back to. Let me share what I observed, and then let's talk about where our views overlap and where they don't.*

Notice what this opening does. It signals immediately that this is a two-way conversation, not a briefing. It honors the employee's experience. It creates space for something the manager did not expect. And it gives the manager one more data point before they present their own assessment: how closely the employee's self-perception aligns with what is in the document. That alignment, or misalignment, shapes everything that follows.

# Structuring the 45-Minute Review Meeting

A well-structured review meeting has a predictable arc that the employee can follow and the manager can navigate. The following framework works for reviews at every performance level, from exceptional to improvement-needed, because the structure is consistent even when the content varies significantly.

<table>
<tr><td>0–5 min<br><br>before</td><td>Opening and Employee Self-Assessment<br><br>• Welcome and brief acknowledgment of the purpose of the meeting<br><br>• Ask the employee to go first: "Walk me through your year"<br><br>• Listen without interrupting. Take brief notes.<br><br>• Close with: "Thank you. Let me share what I observed, and then let's discuss where we see things similarly and where we see things differently."</td></tr>
</table>

<table>
<tr><td>5–20 min<br><br>before</td><td>The Review: Strengths and Accomplishments First<br><br>• Present the review document section by section; do not read it verbatim<br><br>• Begin with the strongest areas and the most significant accomplishments<br><br>• Cite specific evidence for every rating, including what you observed and what the metrics showed<br><br>• Pause after each major section to ask: 'What resonates here? What do you see differently?'<br><br>• Do not rush past the positive sections to get to the developmental ones; let the recognition land</td></tr>
</table>

<table>
<tr><td>20–35 min<br><br>before</td><td>Developmental Feedback and the Hard Conversations<br><br>• Present developmental feedback directly; do not soften it into meaninglessness<br><br>• Name the specific behavior or gap, the impact, and the expectation going forward<br><br>• Allow silence after delivering difficult feedback; resist the urge to fill it immediately<br><br>• If the employee pushes back, hear them fully before responding; the next section of this book covers handling pushback with specific guidance<br><br>• End each developmental section with the forward-facing statement: 'Here is what I need to see in the coming year'</td></tr>
</table>

<table>
<tr><td>35–45 min<br><br>before</td><td>Goals, Development, and Closing<br><br>• Transition from assessment to forward focus: 'Let's talk about where we go from here'<br><br>• Review or introduce the goals for the coming year, briefly rather than exhaustively (the goal-setting meeting is a separate conversation)<br><br>• Name the one or two development priorities for the year ahead<br><br>• Close with your "one thing": the message you most want this person to carry out of the room<br><br>• Ask the employee: 'What questions do you have? What do you want to make sure I understand?'<br><br>• Confirm the next check-in date before the meeting ends</td></tr>
</table>

One structural note: the positive sections come first. Always. Not because you are softening the employee up for bad news, but because strengths deserve genuine recognition before development areas are introduced. An employee who feels seen in their accomplishments is more open to honest developmental feedback than one who spends the first twenty minutes of the meeting waiting for the other shoe to drop. Sequence matters.

## How to Handle the Employee Self-Review

Many organizations require employees to complete a self-review as part of the annual cycle. This is one of the most valuable but most underused elements of the process. Most managers collect the self-review, file it, and proceed as if it had not been submitted. That is a missed opportunity on every level.

The self-review surfaces three things that are very difficult to learn any other way: how the employee perceives their own performance, what they value about their work, and where they feel they have not been seen or recognized. Even when the self-review is poorly written or self-serving, it contains information. Read it carefully before the meeting. Come with questions.

### What to Look for in the Self-Review

- Where does the employee rate themselves significantly higher than you rated them? That gap is worth a conversation, not something to dismiss or override without discussion.
- Where does the employee rate themselves lower than you did? That tells you something about their confidence, their self-awareness, or their understanding of their own impact.
- What accomplishments did they highlight that you had not noted? If they took credit for something you were unaware of, find out whether it is accurate, and acknowledge it if it is.
- What language did they use to describe themselves and their work? The words an employee chooses in a self-review reveal how they think about their role and their identity at work.
- What did they not mention? Silence in a self-review is sometimes as informative as what was written.

## If Your Organization Does Not Use Formal Self-Reviews

Ask anyway. Before presenting your review, ask the employee to spend three to five minutes telling you how they would assess their own year. You do not need a formal document to capture the value of the self-assessment. You need the employee to speak first. The conversation that starts with the employee's own words is consistently richer than the one that starts with the manager's.

# The Complete Pre-Meeting Preparation Checklist

Use this checklist before every review meeting. The items are sequenced by when they should be completed, not all on the morning of the meeting.

## One Week Before the Meeting

- Review is written, self-audited, and finalized; rating matches narrative
- HR has reviewed the document if that is your organization's process
- You have read the review as if you are the employee, from their perspective, not yours
- You have identified the three moments most likely to create friction in this conversation
- You have prepared a brief, honest response for each of those three moments
- You know your "one thing": the message you most want this employee to leave with
- The meeting is scheduled with enough time: 45 to 60 minutes minimum
- The location is confirmed: a private room, not an open space

## 48 Hours Before the Meeting

- The review document has been sent to the employee
- The self-review form has been sent to the employee with a request to complete it
- The meeting agenda has been shared so the employee knows what to expect
- If there is a significant surprise in the review, such as a rating drop or a formal concern, you have given the employee a verbal heads-up so the document is not the first signal

## The Morning of the Meeting

- You have read the employee's completed self-review and noted where your views align and differ
- You have printed two copies of the review, one for you and one for the employee
- The room is set: private, comfortable, water available, no distractions
- Your phone is set to silent or out of reach for the duration of the meeting
- You have blocked time after the meeting; do not schedule something immediately after. You may need it.
- You have taken two minutes to center yourself: the one thing you respect about this employee, your 'one thing' message, your commitment to listen

**Before You Walk In**

- You are not running late; you are in the room before the employee arrives

- The review document is open and accessible, not buried in a stack of other papers

- You know how you are opening the meeting: employee goes first

- You are genuinely present: not mentally in the next meeting, not reviewing emails, not rehearsing a script

> *The checklist is not bureaucracy. It is professionalism. The manager who completes every item on this list before sitting down with an employee has done something most managers never do: they have treated the performance review conversation with the seriousness it deserves. The employee will feel the difference within the first five minutes. So will you.*

## Special Preparation for Difficult Review Situations

Most review meetings follow a predictable arc. But some require additional preparation because the content or context is more complex. Here are the three situations that most commonly catch first-time managers underprepared.

### Delivering a Rating Decrease

When an employee is receiving a lower rating than the prior year, the preparation requirement increases significantly. Before the meeting, you must be able to answer three questions without hesitation: What specifically changed this year? What documentation supports that change? And was the employee given feedback during the year that this was happening, or is this the first time they are hearing it?

If the answer to the last question is that this is the first time they are hearing it, the review conversation will be a very difficult one, and arguably an unfair one. A rating decrease that was not foreshadowed by coaching conversations during the year is a failure of management, not just a performance outcome. Own that, if it applies. Learn from it for next year.

### Delivering a Review That Precedes a Performance Improvement Plan

If this review is the last documented step before a formal PIP, HR must be involved in the preparation. Do not walk into this meeting without having reviewed the language with HR, confirmed the next steps in the process, and prepared a clear statement of what the improvement expectation is and what the timeline looks like. The employee deserves to leave this meeting with absolute clarity about where they stand and what happens next. Ambiguity in this conversation is not kindness; it is a failure to prepare.

## Delivering an Exceptional Review to an Employee Who Is Still Underpaid

This is one of the most uncomfortable situations in performance management, and it happens more often than most organizations acknowledge. The employee has had an outstanding year. The review reflects that. And the compensation adjustment that follows does not match the performance. Before this conversation, know what you can and cannot say about compensation. Know what advocacy you have done on the employee's behalf. Be honest about the gap between the recognition the review reflects and the financial outcome they will receive. An employee who is told they are exceptional and then receives a 2% increase will remember both numbers. Handle the conversation with the honesty it deserves.

### Chapter 12 — Key Takeaways

- Most managers spend weeks writing the review and twenty minutes preparing for the conversation. That ratio needs to reverse.
- Read your own review from the employee's perspective before the meeting. Anticipate the three moments of friction. Know your 'one thing.'
- Send the review 24 to 48 hours in advance. Invite questions and disagreement explicitly. Do not ask for a signature before the conversation.
- Set the room deliberately: private, face to face, no distractions, no phone, enough time. The environment is part of the message.
- The most forgotten step: ask the employee to go first. Let them walk you through their year before you present your assessment. What they say will shape the conversation you need to have.
- Structure the 45-minute meeting: employee self-assessment first, then strengths and accomplishments, then developmental feedback, then goals and close.
- For difficult situations, including rating decreases, pre-PIP conversations, and exceptional reviews paired with inadequate compensation, the preparation requirement increases. Loop in HR. Know what you will say before you walk in.
- Use the preparation checklist. Every item on it represents something that, if missed, will show up in the room.

# Chapter 13

# Handling the Reactions You Didn't Expect

*An employee who reacts strongly to a performance review is not a problem to be managed. They are a human being responding to information that affects how they see themselves and their future. The manager who understands that distinction, and who does not mistake the reaction for the final response, is the manager who gets the outcome the review was designed to produce.*

Before we talk about any specific reaction, we need to establish something that will reframe everything that follows: a strong reaction in a performance review meeting is almost always a sign that the feedback landed. Not that it went wrong. The employee who erupts, or shuts down, or pushes back hard, or cries: that employee heard you. They are taking what you said seriously. That is, in an important sense, the goal.

The employee who nods pleasantly through the entire review, agrees with everything, signs the form, and leaves is often the more concerning outcome. Because what comes next with that employee is silence: no change in behavior, no engagement with the development plan, no evidence that the conversation produced any impact at all. The emotional flatness that feels like a successful review is sometimes just deference. And deference is not the same as growth.

This chapter covers the human mechanics of how people receive difficult feedback, the five most common reactions you will encounter, how to respond to each without escalating or capitulating, and, most importantly, why the conversation you have 48 to 72 hours after the review often matters more than the conversation you had in the room.

## The First Principle: Reactions Are Mitigated Before the Meeting Begins

The single most effective thing a manager can do to prevent difficult reactions in a performance review meeting is to ensure that nothing in the review is a genuine surprise. Not the rating, not the developmental feedback, not the performance gap. Not the goal that was missed.

An employee who is honestly surprised by something documented in their annual review has not been managed well throughout the year. They have been left to perform without feedback, to drift without correction, to assume, sometimes accurately and sometimes not, that things were fine. And then December arrives and the document tells a different story than the one they have been living for eleven months. That surprise is not a property of the review. It is the bill coming due for a year of withheld feedback.

*An associate should never be honestly surprised by anything documented in their annual review. If they are, the problem is not in December. It is in the eleven months that preceded it. The review is a summary of conversations that have already happened, not the first delivery of information the employee has never heard. If it functions as the latter, something fundamental went wrong.*

This is why real-time feedback throughout the year is not just a best practice; it is the primary mechanism of reaction mitigation. Every coaching conversation during the year is a deposit into an account that pays out in December. The manager who has been consistent, honest, and specific with feedback throughout the year walks into the review meeting with an employee who already knows where they stand. The conversation confirms what has been said. It does not reveal what has been withheld.

| Frequency of Real-Time Feedback | Likely Employee Reaction at Review Time | Risk Level for the Manager |
| --- | --- | --- |
| Consistent throughout the year: regular one-on-ones, timely coaching, documented conversations | Familiar with the feedback. May still have feelings about the rating, but nothing in the document is new information. | Low: the conversation is a confirmation, not a revelation |
| Occasional: some feedback conversations, but inconsistent and not always documented | Partially prepared. May push back on specifics they disagree with. Some elements may feel like surprises. | Moderate: expect some pushback, manage it with documentation |
| Rare or absent: little to no formal feedback conversation during the year | Genuinely surprised by negative assessments. Likely to feel blindsided, betrayed, or treated unfairly. Reactions will be stronger and less manageable. | High: employee has legitimate grievance. The manager faces both legal and relational risk. |
| Only for problems: feedback delivered only when something goes wrong, never for recognition or development | Conditioned to associate feedback with failure. Will arrive at the review in a defensive posture before a word is spoken. | Elevated: the emotional climate is already adversarial before the meeting begins |

The table above is worth sitting with. The manager in the high-risk row did not create that risk in the review meeting. They created it in February, and April, and July, every time a coaching conversation was avoided because it felt uncomfortable. The review meeting is where that accumulated avoidance becomes visible, and where the employee, fairly or not, experiences the cost of it.

# The 48-72 Hour Window — Understanding How Humans Actually Process Feedback

This is the most important concept in this chapter, and possibly the most important concept in this book, for a manager navigating a difficult review conversation. It is grounded in how human beings actually respond to corrective or constructive feedback, not how we wish they would respond, and not how management training programs suggest they should respond. How they actually do.

When a person receives feedback that challenges their self-concept, feedback that says, in effect, "the story you have been telling yourself about your performance is not the story I have been observing," the immediate response is almost always externalization. The mind pushes the discomfort outward. It finds a reason the feedback is wrong, unfair, incomplete, or motivated by something other than honest assessment. This is not weakness or immaturity. It is biology. It is the psychological immune system doing exactly what it was designed to do: protecting the self from information that threatens it.

The manager who does not understand this will make a predictable mistake: they will interpret the externalization as the employee's final position. They will escalate, defend, or double down in response to the pushback. And in doing so, they will convert a temporary emotional response into a hardened position, because now the employee is defending themselves not just against the feedback, but against the manager's escalation.

The manager who does understand this will do something different. They will hold their ground calmly. They will not chase the emotion. They will allow the reaction without amplifying it. And they will give the employee time.

## The Five Phases of Feedback Processing

What follows is not a clinical model. It is what actually happens, observed across decades of delivering performance feedback at every level of an organization, to people at every performance level. The timeline varies by individual, by the severity of the feedback, and by the quality of the manager-employee relationship. But the arc is consistent.

| | |
|---|---|
| **0–2 Hours**<br>*Externalization* | The immediate, instinctive response. Defensiveness, denial, anger, silence, or withdrawal. This is biology, not character. Do not judge it. Do not escalate into it. |
| **2–24 Hours**<br>*Processing* | The employee begins replaying the conversation. They may seek validation from peers. They may cycle between anger and acceptance. They are not yet ready to commit to change. |
| **24–48 Hours**<br>*Transition* | The emotional charge begins to diminish. The employee starts to separate the feedback from the feelings it produced. Rationalization may occur, but so does the beginning of honest reflection. |
| **48–72 Hours**<br>*Internalization* | The employee begins to own the feedback. Not necessarily to agree with all of it, but to engage with it honestly. This is when the productive conversation can happen. Not before. |
| **72+ Hours**<br>*Integration* | The feedback becomes part of how the employee understands themselves and their work. Behavior change, if it is going to happen, begins here. This is the outcome the review was designed to produce. |

The critical insight for the manager is this: the conversation that happens in the room, in that first two hours, is rarely the conversation that determines the outcome. The outcome is determined by what the employee decides when they are alone with the feedback, in the processing and transition phases, when the emotion has quieted and the thinking begins.

Your job in the room is not to win the argument. It is not to convince the employee that you are right before they leave. It is to deliver the feedback with enough clarity and specificity that when the employee enters the internalization phase, 48 to 72 hours later, they have something accurate and honest to work with. Muddled feedback, softened into meaninglessness to avoid a reaction, gives the employee nothing to internalize. Specific, documented, evidence-based feedback gives them something real to reckon with. Give them something real.

*The 48-72- hours window is not a management technique. It is a human fact. We externalize first. We internalize later. The manager who knows this does not panic when the immediate reaction is defensive or emotional. They hold the space, hold the message, and give the employee time to arrive at the place where the real work begins. Patience in those 72 hours is not passivity. It is wisdom.*

## What to Do in the 48-72 Hours After a Difficult Review

Give the employee space. Do not follow up within 24 hours to check in on how they are feeling about the review. Do not interpret their silence as escalating concern. Do not send emails that revisit or re-explain what was said in the meeting. The employee is processing. That processing requires room. Give it to them.

After 48 to 72 hours, not before, a brief, low-key follow-up is appropriate. Not a formal meeting. Not an interrogation about whether they have "come around." A simple check-in: "I wanted to follow up on our conversation. Do you have any questions, or anything you want to talk through?" That opening is an invitation, not a demand. The employee who is ready to engage will. The employee who needs more time will say so. Both responses are acceptable.

<table>
<tr><td>Sample 48-72 Hour Follow-Up — Manager to Employee</td></tr>
</table>

**MANAGER: I wanted to check in after our review conversation earlier this week.**
*I know that was a lot to process, and I don't expect you to have resolved everything in a few days.*

*Is there anything from our conversation you want to revisit, push back on, or talk through? I'm not looking for agreement; I'm looking to make sure you have what you need to move forward.*

**EMPLOYEE: [Space for the employee to respond; the manager listens without**
*defending or re-litigating the review.]*

**MANAGER: [After the employee speaks] I appreciate you sharing that. What I**
*want you to know is that the feedback I gave you came from a genuine belief that you are capable of more than you showed this year, and that's what the next twelve months are about. I'm not interested in the rating. I'm interested in what happens next.*

Notice what that closing line does. It redirects from the review, which is the past, to the year ahead, which is where both parties actually have power. The review is a record. The year ahead is an opportunity. The manager who makes that transition explicitly gives the employee something to move toward rather than something to argue about.

# The Five Reactions: What They Look Like and How to Handle Each

The following five reactions cover the vast majority of what a first-time manager will encounter in a review meeting. None of them are evidence that the review process failed. All of them are manageable with the right response. The wrong response, in every case, is escalation.

## Reaction 1: The Pushback — 'I Disagree with This Rating'

### What It Looks Like

The employee disputes a specific rating or piece of feedback directly. They present counter-evidence, cite examples the manager did not include, or challenge the fairness of the standard applied. The tone may be calm and professional or heated and urgent.

### Why It Happens

This is often the healthiest reaction in the room. An employee who pushes back is engaged, self-aware, and willing to advocate for themselves. The employee who accepts everything silently may be processing, or they may be disengaged. The pusher-back is neither passive nor indifferent.

### ✗ The Wrong Response

Becoming defensive about your own assessment. Repeating yourself more loudly. Shutting down the conversation with authority: "This is my rating and it stands." All of these responses convert a productive dialogue into a power struggle, one you may win in the room and lose everywhere else.

### ✔ The Right Response

Hear the counter-argument fully before responding. Ask: "Tell me more about what you are pointing to." If the evidence they raise is legitimate and you were not aware of it, acknowledge it, and if warranted, adjust. If your rating is well-documented and accurate, hold it calmly: "I hear your perspective and I want you to know I considered it. The documentation I have supports where I landed. I am open to continuing this conversation, but the rating reflects what I observed." Then stop. Do not re-litigate every piece of evidence. State your position, name your documentation, and allow the employee to process.

## Reaction 2: The Emotional Response — Tears, Anger, or Distress

### What It Looks Like

The employee becomes visibly emotional. This may present as tears, a raised voice, visible frustration, or a sudden change in body language: turned away, arms crossed, jaw tight. The emotion may be proportionate to the severity of the feedback or it may seem disproportionate to what was just said.

### Why It Happens

Emotions in a review meeting are not irrational. They are a signal that the feedback matters to this person, that their work is not separate from their identity, and that information about their performance is information about themselves. The most committed employees are often the most emotional in reviews, because they care the most.

### ✗ The Wrong Response

Rushing to comfort in a way that softens or retracts the feedback. "Oh, I didn't mean it was that bad" undoes the honest message you just delivered. Conversely, pressing forward as if the emotion is not happening, continuing to read from the review document while the employee is visibly distressed, signals a complete absence of human awareness.

### ✔ The Right Response

Pause. Name what you see, not what you assume: "I can see this is landing hard. Let's take a moment." Then be quiet. Silence is not awkward; it is respectful. Give the employee time to collect themselves without commentary from you. When they are ready to continue, ask: "Do you want to keep going, or would it help to take a short break and pick this back up in a few minutes?" Both options are valid. The employee decides.

## Reaction 3: The Silence — Shutdown or Withdrawal

### What It Looks Like

The employee goes quiet. They stop engaging, stop asking questions, give minimal responses. They may nod without speaking. The energy in the room drops. This can easily be misread as acceptance when it is often something quite different.

### Why It Happens

Silence during a difficult review usually means one of three things: the employee is overwhelmed and has shut down to protect themselves; they are angry and have decided to give nothing rather than say something they will regret; or they have withdrawn into a private decision about the conversation that they are not yet willing to share. None of these is indifference.

### ✗ The Wrong Response

Interpreting silence as agreement and moving forward as if everything is fine. "Great, so we're aligned on this; let's talk about goals." The employee who was silently furious is now silently furious and unheard. Alternatively, pressing the employee to speak: "I need you to respond to this. What are you thinking?" Demanding emotional openness from someone who has just shut down is a guarantee of a worse outcome.

### ✔ The Right Response

Name the silence without judgment and invite, rather than demand, a response: "I notice you've gotten quiet. That's completely okay. You don't have to have a response right now. What I want you to know is that I mean what I said, and I believe it came from a fair assessment. If you want to continue today, I'm here. If you need some time to think about it first, we can schedule a follow-up conversation in a couple of days." Then wait. Do not fill the space. The employee will signal what they need.

## Reaction 4: The Counter-Complaint — 'You Never Told Me'

### What It Looks Like

The employee's pushback is not about the rating itself but about the process: "This is the first time I'm hearing this." "If this was a problem, why didn't you say something sooner?" "I feel blindsided." They are not disputing the evidence; they are disputing the fairness of the timing.

### Why It Happens

This reaction is often the most accurate one in the room. If the employee truly was not told about a performance gap until December, if the regular coaching conversations throughout the year did not happen, their complaint is legitimate. The manager in this situation is not being unfairly challenged. They are being held accountable for a failure of management that preceded the review.

### ✗ The Wrong Response

Becoming defensive about the timing. "I gave you feedback throughout the year." If the documentation does not support this, the statement is not credible and the employee knows it. Deflecting: "That's not the point of today's conversation." It is absolutely the point, because the employee cannot act on feedback they were never given.

### ✔ The Right Response

If the complaint is legitimate: own it. "You're right that I could have been more explicit about this earlier in the year. That's on me. What I want to focus on now is making sure you have everything you need going forward, starting with this conversation." If the complaint is not accurate and you have documentation of prior feedback: "I understand it may not have felt like formal feedback, but we did discuss this on [specific dates]. Those conversations are what informed this assessment. I hear that the impact wasn't what I intended, and I'll be more deliberate going forward." Then move forward.

## Reaction 5: The Complete Surprise — 'I Had No Idea'

### What It Looks Like

The employee appears genuinely shocked. Not by the rating level, but by specific content in the review: a behavioral observation, a documented gap, a pattern the manager has named. They may say "I didn't know this was how you saw me" or "I thought I was doing well in this area." Their reaction is not defensive; it is genuinely disoriented.

### Why It Happens

If this reaction is authentic, and experienced managers can usually tell, it carries the most diagnostic information of any reaction in this list. An employee who is genuinely surprised by specific feedback is telling you that either: the feedback was not communicated during the year as clearly as the manager believed; the employee's self-perception gap is significant and is itself a development issue; or the manager's observation is accurate but the employee has been operating without awareness of the impact of their behavior.

### ✗ The Wrong Response

Treating the surprise as manipulation or performance. 'I don't believe you didn't know about this.' If the employee is being honest, this response destroys trust instantly. Alternatively, retracting or softening the feedback because the employee's surprise made you doubt yourself. If your documentation is solid, the surprise does not change the accuracy of the assessment.

### ✔ The Right Response

Treat the surprise as information, not as resistance: "I hear that this is landing differently than I expected. Let me walk you through what I observed and why it formed this part of my assessment." Then provide the specific evidence, including the dates, the behaviors, and the impact. The goal is not to convince the employee in the room. It is to give them accurate, specific information to carry into the internalization window. The processing will happen later. Your job now is to make sure what they are processing is real.

# The Most Damaging Mistakes Managers Make in the Room

Each of the following mistakes is common. Each is understandable; they come from the same instinct to reduce discomfort and resolve tension quickly. And each one damages either the integrity of the review, the manager-employee relationship, or both.

## Softening the Feedback Mid-Conversation

The employee reacts. The manager, feeling the discomfort of having caused distress, begins to walk back what they said. "What I really meant was..." "It's not as bad as it may have sounded..." "I think I phrased that wrong..." Each of these phrases signals to the employee that if they react strongly enough, the manager will retreat. This is one of the fastest ways to undermine your own credibility and make every future feedback conversation harder. If the feedback was accurate before the reaction, it is accurate after the reaction. Hold it.

## Escalating Into the Reaction

The employee becomes heated. The manager becomes heated in response. The conversation becomes an argument. This is the outcome most catastrophically misaligned with the purpose of a performance review. A review meeting that escalates into a confrontation does not produce development; it produces grievances, distrust, and occasionally formal complaints. The rule is absolute: whatever the employee's temperature in the room, the manager's temperature stays constant. Calm is not weakness. It is leadership.

## Asking for Immediate Agreement

At the end of a difficult review, the manager, wanting resolution, asks: "Do you understand?" or "Can you commit to this?" or worst of all, "Do you agree with this assessment?" These questions do not produce genuine agreement. They produce compliance in the moment, an employee who says yes because saying no feels unsafe, and resentment afterward. Do not ask the employee to agree with the review in the room. Ask them to hear it, to think about it, and to come back with their questions. Agreement, if it comes, arrives in the internalization window. Not before.

## Making It Personal

The performance review is an assessment of work, behaviors, and results. The moment it becomes an assessment of character, even implicitly, the conversation has crossed a line that is very difficult to walk back. "You don't seem to care about this." "I've been doing this a long time and I can tell when someone is coasting." "Frankly, I expected more from you at this point." These statements may feel honest in the moment. They are not documentation. They are not specific. And they make the employee feel personally attacked rather than professionally assessed, which closes them to the feedback rather than opening them to it.

## Allowing the Conversation to End Without Clarity on Next Steps

A difficult review meeting that ends in tension, or tears, or extended silence, and then simply ends, is a conversation without a landing. Before the employee leaves the room, regardless of how the meeting went, confirm two things: when the follow-up conversation will happen, and that the door is open for questions in the meantime. The meeting may have been hard. The relationship is still intact. Those two confirmations signal both.

> *The standard for a manager in a difficult review conversation is not warmth. It is not authority. It is not efficiency. It is steadiness. The manager who remains steady, who does not chase the emotion, retreat from the feedback, or escalate the tension, is the manager whose employee returns 72 hours later ready to do something with what they heard. Steadiness is the thing. Practice it before you need it.*

# When to Involve HR During or After the Review Conversation

Most difficult review conversations, even the very uncomfortable ones, do not require HR intervention. They require a manager who is prepared, steady, and patient. But certain situations cross a threshold where HR involvement is not optional. Know the line before you are standing on it.

## Involve HR Immediately If:

- The employee makes a formal complaint or allegation during the review meeting, including discrimination, harassment, retaliation, or any claim that the review itself is retaliatory
- The employee becomes threatening, whether toward themselves or toward others, even vaguely or implicitly
- The employee discloses a medical or mental health situation that may have legal implications for how the review is handled
- The employee states they intend to formally dispute the review through a grievance process
- The conversation becomes so heated that continuing it safely is not possible

**Involve HR Within 24 Hours If:**

- The employee made claims during the review that you were not able to address in the moment and that may have merit
- The review conversation revealed information that changes your understanding of the employee's situation in ways that may affect the assessment
- You said something during the meeting that you are uncertain about from a legal standpoint
- The employee left the meeting in a state that concerns you for their wellbeing

When in doubt, call HR. Not because you are in trouble, but because their job is to help you navigate exactly these situations, and calling them the day after a difficult conversation is significantly better than calling them three weeks later when a formal complaint has already been filed.

## The 72-Hour Follow-Up — Where the Real Outcome Lives

By now the principle is clear: the conversation in the room is not the final word. The employee is still in the externalization and processing phases during the meeting itself. The internalization, the genuine reckoning with the feedback and the decision about what to do with it, happens in the 72 hours that follow. Which means the most important performance management conversation you have this year may be the one you have three days after the formal review.

That follow-up conversation has a different quality than the review meeting. The stakes feel lower. The formality is gone. The document is not in front of both of you. And if you have given the employee time to process without pressure, they will often come to that second conversation having done something the first conversation could not produce: they will have made a decision about the feedback. They may have decided you were right. They may have decided you were partially right. They may have decided they want to push back further. All of those positions are more useful than the defensive or compliant performance of agreement that often ends the first meeting.

**What to Listen for in the Follow-Up**

- The employee who says 'I've been thinking about what you said and I think I understand it better now' has entered internalization. This is the most productive place the conversation can go. Meet them there with specificity about the path forward.
- The employee who says 'I still don't agree with parts of it but I'm ready to move forward' has reached a functional resolution. They do not have to agree. They have to be able to engage with the work. Accept this as a good outcome.
- The employee who says 'I've been thinking about it and I want to talk about [specific element]' is still processing actively. This is healthy. Have the conversation with patience and without defensiveness.

- The employee who says nothing, who has not visibly processed and has not come back with questions or reflection, may need a direct check-in: "I want to make sure we're in a productive place. Can you tell me where you are with our conversation?" Do not let the silence harden into distance.

## Chapter 13 — Key Takeaways

- A strong reaction to a review is not a sign that something went wrong. It is usually a sign that the feedback landed. The employee who nods and agrees with everything is sometimes the more concerning outcome.
- The most powerful reaction mitigation happens before the meeting, through real-time feedback throughout the year. An employee who is honestly surprised by their annual review has not been managed well during the year preceding it.
- The Externalization-to-Internalization Arc: humans externalize defensive reactions first, always. Internalization, the honest reckoning with feedback, happens 48 to 72 hours later. The manager who knows this does not mistake the first reaction for the final one.
- In the room: deliver the feedback with clarity and hold it steadily. Do not soften it under pressure. Do not escalate into the emotion. Do not ask for immediate agreement. Give the employee something accurate to carry into the internalization window.
- The five reactions, pushback, emotion, silence, counter-complaint, and genuine surprise, each require a different response. All five require the same manager quality: steadiness.
- The most damaging mistakes are softening mid-conversation, escalating into the reaction, asking for immediate agreement, making it personal, and allowing the meeting to end without clarity on next steps.
- The 72-hour follow-up is where the real outcome lives. After the employee has had time to process, return to the conversation. Listen for where they have landed. That conversation determines what actually happens next.

# Chapter 14

# The Follow-Up That Most Managers Skip

*The signed review is not the finish line. It is the starting line. Everything that was written, discussed, and agreed to in December is worth exactly what happens in January, February, and March. The manager who follows up turns a conversation into a commitment. The manager who doesn't turns it into paperwork.*

Here is the pattern that plays out in organizations every year, reliably enough to be called a cycle: Review season arrives. Managers write, deliver, and file performance reviews. Employees read them, react to them, sign them. And then the reviews disappear, vanishing into HR systems, desk drawers, and the gap between intention and action, and are not looked at again until the following December, when the manager sits down to write the next one and discovers they cannot remember what they committed to twelve months ago.

The development plan that was written in the review is not followed. The goal adjustment that was agreed to is not tracked. The coaching commitment the manager made in the closing statement is not kept. The employee who left the review meeting believing something was going to change discovers, slowly and then all at once, that nothing did. And they draw the only reasonable conclusion available to them: the review is theater. The real work is elsewhere.

This chapter is about breaking that pattern. It covers what must happen in the first 30 days after the review, how to activate a development plan rather than file it, how to check in on progress without micromanaging, and how the follow-up habits of a manager determine whether the performance cycle produces real change or just annual documentation. The tools in this chapter are simple. The discipline to use them consistently is what most managers lack, and what the managers whose employees grow fastest have built.

## The First 30 Days — When the Review Either Lives or Dies

Research on behavioral change is consistent on one point: commitments made without near-term follow-up do not produce behavior change. They produce good intentions that fade under the weight of daily demands. The first 30 days after a performance review are the highest-leverage window in the entire cycle. What happens in those 30 days determines whether the review was a professional development event or an annual compliance exercise.

There are five specific actions that must happen in the first 30 days. Not eventually. Within 30 days.

## Action 1: Activate the Development Plan

The development plan written in Section 5 of the annual review template is not active until something is scheduled. A course with no enrollment date is not a development activity. A mentor relationship with no first meeting date is not a mentor relationship. A stretch assignment with no start date is an intention. Within the first 30 days, every development activity in the plan needs a date attached to it. A calendar entry. A registration confirmation. A first conversation. Something that makes it real.

## Action 2: Schedule the First Progress Check-In

The 30-day check-in is not a mini-review. It is a focused fifteen-to-twenty-minute conversation with one purpose: to confirm that the employee knows what they are working on and has what they need to start working on it. It is the bridge between the review conversation and the year. Without it, the employee is left to interpret the review on their own, and their interpretation may or may not match what was intended.

## Action 3: Ensure Goals Are Visible and Tracked

The goals established in the review, or in the separate goal-setting meeting that follows, should be somewhere the employee and manager can both see them and reference them throughout the year. Not buried in an HR system that requires three logins to access. Visible. In the one-on-one agenda, in the shared document, in the folder on the manager's desktop. If the goals are not visible, they are not being managed. They are being stored.

## Action 4: Send the Commitment Email

Within one week of the review meeting, send the employee a brief email summarizing the commitments made on both sides. Not the entire review. Just the commitments. What you agreed to provide. What the employee agreed to pursue. What the first milestone is. This email is the paper trail of accountability that makes both parties' commitments real. Without it, everything that was said in the room is subject to memory, which is not a reliable system.

## Action 5: Begin Documenting the New Year

The documentation habit does not restart in June. It restarts in January. The file for this employee's next review cycle opens the week the current review closes. The first entry is the review itself: what was said, what was agreed to, what the rating was. The second entry is the first meaningful observation of the new year. Build the file from day one.

## The 30-Day Check-In Template

The 30-day check-in should take fifteen minutes. It is not a formal meeting. It is a structured conversation, whether over coffee, a quick walk, or five minutes at the end of a one-on-one, that serves one purpose: to ensure the employee is oriented and moving. Use the following four questions as your guide.

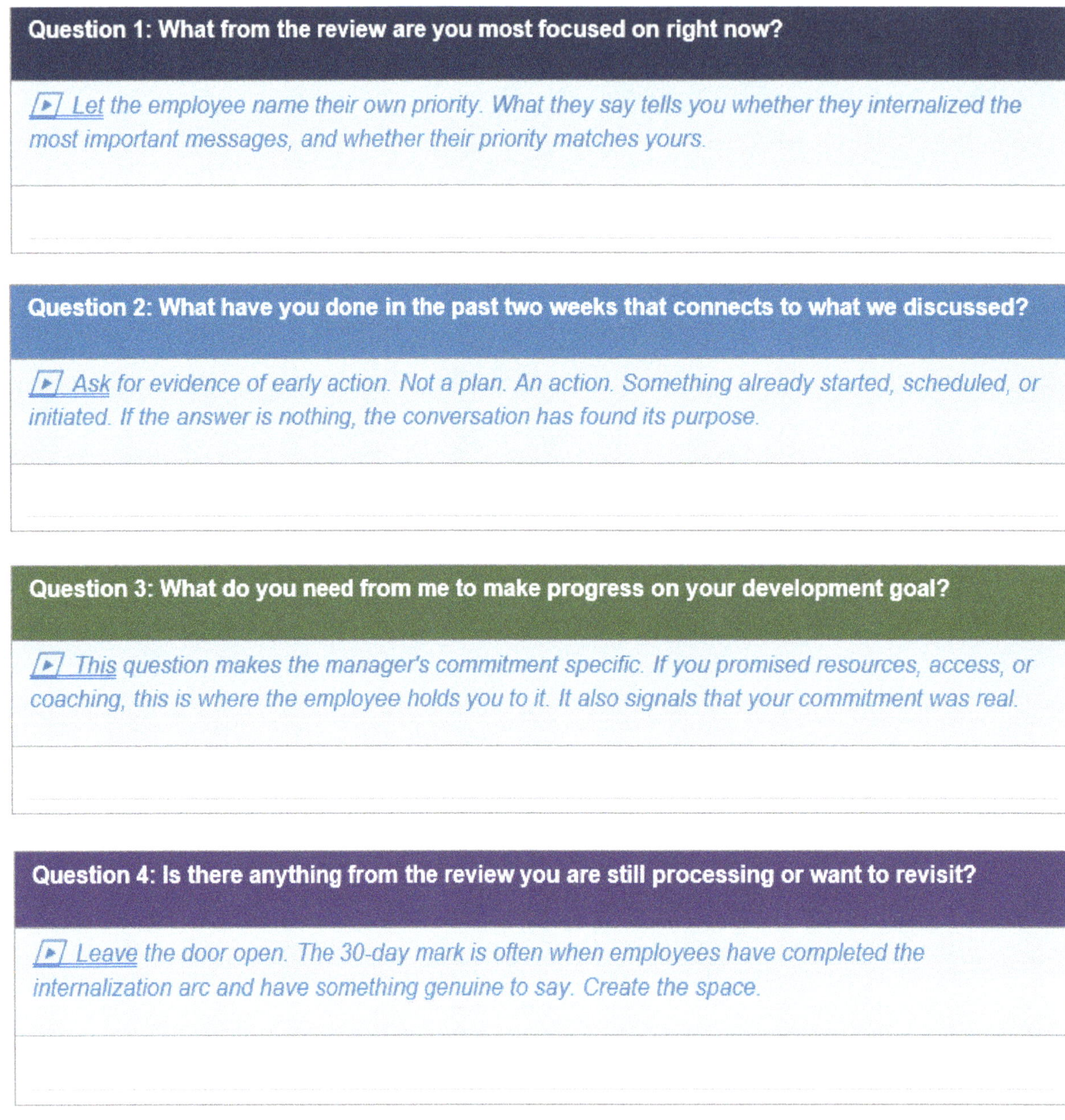

**Question 1: What from the review are you most focused on right now?**

*▶ Let the employee name their own priority. What they say tells you whether they internalized the most important messages, and whether their priority matches yours.*

**Question 2: What have you done in the past two weeks that connects to what we discussed?**

*▶ Ask for evidence of early action. Not a plan. An action. Something already started, scheduled, or initiated. If the answer is nothing, the conversation has found its purpose.*

**Question 3: What do you need from me to make progress on your development goal?**

*▶ This question makes the manager's commitment specific. If you promised resources, access, or coaching, this is where the employee holds you to it. It also signals that your commitment was real.*

**Question 4: Is there anything from the review you are still processing or want to revisit?**

*▶ Leave the door open. The 30-day mark is often when employees have completed the internalization arc and have something genuine to say. Create the space.*

## The Commitment Email — Making Both Sides Accountable

This is the most skipped step in the entire performance cycle, and the most consequential. The commitment email transforms verbal agreements into written ones. It creates a shared record that both parties can reference. And it signals to the employee that the manager takes the review seriously enough to document what they promised.

**Sample Commitment Email — Within One Week of the Review Meeting**

*Subject: Review Follow-Up: Commitments and Next Steps for [Employee Name]*

*Hi [Name],*

*I wanted to follow up on our review conversation with a brief summary of what we agreed to so we have a shared reference going into the new year.*

*What you committed to:*
*- Complete the Learn On supervisory development course by end of Q1*
*- Lead the agenda for one section of the Q3 manager summit*
*- Bring a written reflection from your first customer ride-along by March 31*

*What I committed to:*
*- Connect you with [Senior Manager Name] for a development conversation by the end of January*
*- Review your IDP progress at our monthly one-on-one*
*- Advocate for your involvement in the Q2 cross-functional project*

*First milestone: A quick check-in on February 1 to confirm the Learn On course is enrolled and scheduled.*

The format above is not a template to follow word for word. It is a model for the three elements every commitment email needs: what the employee committed to, what the manager committed to, and the first specific milestone with a date. The manager's commitments are as important as the employee's. When a manager writes down what they promised, they are held to it by the employee, by the record, and by their own integrity.

# Turning a Development Plan Into an Actual Development Plan

A development plan that exists only in a review document is not a development plan. It is a list of aspirations written to satisfy a form requirement. The difference between a development plan that produces growth and one that produces documentation is activation, the specific steps that convert intention into action and action into evidence.

## The Four-Part Activation Model

For each development goal in the review, apply all four of the following:

- Name the first action, not the goal, the first action. 'Improve strategic thinking' is a goal. 'Enroll in the APICS certification program by January 31' is a first action. The first action must be specific enough that next week you can check whether it happened.
- Name the resource or relationship required. Development rarely happens in isolation. Does this goal require a mentor? A course enrollment? A project assignment? A conversation with a peer in another building? Name it. Then arrange it. Within 30 days.
- Name the evidence milestone. How will you and the employee know, at the mid-year check-in, that progress is real? Not effort. Evidence. 'Has completed the course' is evidence. 'Has been working on it' is not. Evidence must be observable or measurable.
- Name your role as the manager. Development is not something that happens to an employee. It is something that happens with a manager's active support. What will you do to enable this goal? What doors will you open, what feedback will you give, what time will you protect? Write it down. It is a commitment.

*The development plan that gets activated in January gets evidence by June. The development plan that gets filed in December gets rediscovered the following December, unchanged, unmet, and quietly embarrassing for everyone. Activation is everything.*

# The Quarterly Check-In Rhythm — Staying Connected Without Micromanaging

The performance cycle does not go dormant after January. It runs continuously, fed by the documentation habit covered earlier in this book, maintained by the mid-year check-in, and kept alive by the quarterly rhythm described here. The quarterly check-in is not a review. It is a calibration. A fifteen to twenty-minute conversation, four times a year, that keeps both the manager and the employee oriented to the same picture of where things stand.

## Q1 Check-In — January to March

Focus: activation and early momentum. Are development activities started? Are goals being pursued? Is there anything in the environment that has changed since the review that affects the goals or the plan? The Q1 check-in is the earliest opportunity to course-correct before a small drift becomes a significant gap.

## Q2 Check-In — April to June (Mid-Year)

The formal mid-year check-in is the most significant checkpoint in the cycle outside of the annual review. Goal status is formally assessed. Development progress is reviewed. The second half of the year is oriented.

## Q3 Check-In — July to September

Focus: trajectory and adjustment. Is the employee on a path to year-end success? Are any goals at risk? Are any development goals showing real evidence of growth? The Q3 check-in is when the manager should be looking at the documentation file and asking: if the year ended today, what would the review say? If the answer is concerning, the Q3 check-in is the last meaningful opportunity to change the trajectory before December.

## Q4 Check-In — October to December (Pre-Review)

Focus: preparation and closure. This is the conversation that precedes the review writing, not the review itself. Pull the documentation file. Review the goals. Assess the development evidence. And have an informal conversation with the employee: 'We're getting close to review season. I want to make sure you feel prepared and that nothing in your review is a surprise.' That conversation is both a courtesy and a final check against unwanted surprises.

# The Full Performance Cycle — Closing the Loop

Every tool, template, and technique in this book connects to a single continuous cycle. The cycle has no beginning and no end. It is always running. Here is what that looks like when it is working.

| | |
|---|---|
| **01** January | Goal-setting conversation and development plan activated. Documentation file opened. Commitment email sent. |
| **02** Feb–May | Monthly 15-minute documentation habit maintained. Quarterly Q1 check-in. One-on-ones reference goals and development. Coaching conversations documented. |
| **03** June | Mid-year check-in conducted. Goal status reviewed. Development progress assessed. Second-half direction set. Follow-up email sent. |
| **04** Jul–Sep | Monthly 15 continues. Q3 check-in. Trajectory assessed. Documentation file reviewed for gaps. Any necessary coaching conversations held and documented. |
| **05** Oct–Nov | Pre-review Q4 check-in. Documentation file reviewed and organized. Manager reads the year as a whole. Employee self-review requested. |
| **06** December | Review written from the documentation file, not from memory. Self-review read and incorporated. Review delivered after full preparation. |
| **07** Post-Review | Reactions managed with steadiness. 72-hour follow-up conversation. Commitment email sent. 30-day check-in scheduled. New documentation file opened. Cycle begins again. |

Look at step seven. The cycle does not end with the review meeting. It ends, and begins again, with the follow-up actions that activate the next cycle. The manager who completes step seven is not finishing the year. They are starting the next one with every advantage the current year built.

> *The performance cycle is not an annual event. It is a continuous professional relationship between a manager and an employee, formalized at certain points and informal at others, documented throughout, and always oriented toward the same question: is this person growing, and am I playing my role in that growth? The review is the most visible moment in that relationship. It is not the most important one.*

# What Consistent Follow-Up Builds Over Time

This chapter has been practical: templates, timelines, check-in questions. But the real argument for follow-up is not efficiency. It is something harder to measure and more important to name.

When a manager consistently follows up on what they said in a review, checking in, activating the development plan, tracking the goals, and referencing the conversation months later, they build something that no single review document can create on its own. They build credibility. The employee learns that when this manager writes something down, it matters. That when they make a commitment, they keep it. That the review is not theater. It is the beginning of something real.

That credibility compounds. The employee who has experienced a manager who follows through is more open in the next review. More honest in their self-assessment. More willing to engage with developmental feedback. More likely to stay, because people do not leave jobs that are developing them under a manager who is genuinely invested in their growth.

None of that is produced by a well-written review. All of it is produced by what happens after the review is filed. That is the argument for this chapter. That is the argument for this book. The words matter. The conversation matters. And what you do on the Monday after the review matters most of all.

---

### Chapter 14 — Key Takeaways

• The signed review is the starting line, not the finish line. What happens in the first 30 days determines whether the review produces change or just documentation.

• Five actions within 30 days: activate the development plan, schedule the first check-in, make goals visible and tracked, send the commitment email, and begin documenting the new year.

• The commitment email, sent within one week of the review, names what both parties agreed to and the first milestone. It makes verbal commitments written ones.

• Development plans require activation: a first action with a date, a named resource or relationship, an evidence milestone, and an explicit manager commitment.

• The quarterly check in rhythm (Q1 activation, Q2 mid-year, Q3 trajectory, Q4 pre-review) keeps the performance cycle running continuously rather than annually.

• The full cycle has seven steps. Step seven, the follow-up, is what starts the next cycle. The manager who completes it begins the new year with every advantage the current year built.

• Follow-up builds credibility over time. The employee who experiences a manager who keeps commitments is more open, more honest, and more likely to stay. That outcome is worth every follow-up conversation.

---

**Kevin Baker**

*The conclusion addresses the cumulative effect of doing this work well over a career: what it means for the employees who were developed, the teams that were built, and the manager who built them.*

# Conclusion

# You Are Building Something With Every Review You Write

*The best review I ever received was three paragraphs. It said exactly what I had done, exactly why it mattered, and exactly what the person who wrote it believed I was capable of next. I kept it for fifteen years. I can still tell you what it said.*

That review was not kept because the rating was high, though it was. It was not kept because the language was polished, though it was that too. It was kept because reading it felt like being truly seen by someone who had been paying attention. Who knew the work. Who understood what it had cost. Who believed, specifically and in writing, that what had been built was real and worth building on.

That is what this book has been trying to give you the tools to do. Not to write reviews that satisfy a form requirement. Not to produce documentation that protects the organization from legal exposure, though that matters too. But to write reviews that people keep. That shape how someone understands their own work. That arrive at the right moment in a person's career and say: I see you. I've been watching. Here is what I know.

## What You Have Been Building

Every performance review you write is a contribution to someone's professional biography. Most employees will receive somewhere between thirty and fifty performance reviews across the span of a career. A handful of those reviews, the ones written with care, specificity, and genuine observation, will matter in ways that outlast the rating cycle they came from. They will be referenced in promotion conversations. They will be shared with new managers as context. They will be read on a hard night, when an employee needs to remember that someone once wrote clearly and specifically about why their work was good.

The reviews you write casually, the vague ones, the generic ones, the ones written the week before the deadline from memory, also contribute to someone's professional biography. They contribute by their absence. The employee who receives a review that could have been written for anyone receives a signal: my manager had no idea what I was doing. And that signal, delivered consistently, produces an employee who stops giving their manager something worth watching.

You are building something with every review you write. The question is whether you are building it intentionally.

## The Manager Who Started on the Warehouse Floor

There is a particular authority that comes from having done the work. From having stood on the floor at 5 AM, from having supervised a team through a peak season with half the staff you needed, from having written a disciplinary record for someone you genuinely cared about and having done it because it was the right thing to do. That authority cannot be taught in a classroom or acquired from a framework. It comes from years of showing up and doing the hard thing.

The performance management system works when the people operating it have that authority, when they have earned the right to assess someone else's work because they have done the work themselves, because they understand what it costs and what it produces, because they can look an employee in the eye and deliver honest feedback without flinching because they know what honesty in that moment actually requires.

The first-time manager reading this book is at the beginning of building that authority. This book is not a substitute for the experience. It is a companion to it, a framework that helps you use what you are learning in the field, name what you are observing, and translate both into language that serves the people who report to you.

## The Three Things That Actually Matter

After everything in this book, the templates, the phrase banks, the legal guardrails, the conversation frameworks, the documentation systems, the performance review process comes down to three things. Not thirteen chapters of them. Just three.

1. Pay attention all year. Not just in December, not just when something goes wrong. To all of your people, with enough consistency that when review time arrives you are selecting from evidence rather than reconstructing from memory. The documentation system, the monthly fifteen minutes, the file that grows from January: all of it serves this one thing.

2. Tell the truth. Specifically, fairly, and with enough care that the truth can be received. Not softened into uselessness. Not hardened into cruelty. The honest assessment of someone's work, written by a manager who was present all year, is one of the most useful professional gifts it is possible to give. The vague one, the inflated one, the one that avoids the hard thing: those cost the employee something they can never recover: accurate information about where they stand and what to do about it.

3. Follow through. The commitment email. The 30-day check-in. The development plan with dates attached. The conversation 72 hours after the difficult review when the employee has had time to process and is ready to say something real. The review that is followed up on becomes the beginning of something. The review that is filed becomes the record of something that ended.

> *Pay attention all year. Tell the truth specifically. Follow through on what you said. Everything else in this book is in service of those three things. If you do all three, consistently, for every person who reports to you, you will not just write better reviews. You will build better teams. That is what this was always about.*

## A Final Word on the People

Behind every template in this book is a person. Behind the phrase banks is someone sitting across from a manager waiting to find out if their year mattered. Behind the documentation system is an employee who worked hard and deserves to have that work remembered accurately. Behind the legal guardrails is a human being whose livelihood, self-concept, and professional future are affected by what ends up in the permanent record.

The performance review is a powerful document. It shapes compensation decisions, promotion conversations, succession plans, and termination justifications. It creates the official narrative of someone's professional year. Most people receive somewhere between thirty and fifty of them in a career, and they remember the ones that were written with care, not because they were flattering, but because someone took the time to get it right.

You have that opportunity, every review cycle, for every person on your team. The tools are here. The framework is here. The examples are here, drawn from real reviews written by a manager who believed that doing this work well was worth the time it required.

It is worth the time. Do the work.

# Appendix

# Master Phrase Bank

This phrase bank collects the most useful performance review language from throughout this book, organized by category, performance level, and competency area. All phrases are starting points. Customize each with specific evidence from your documentation file. A phrase without evidence behind it is still a vague review.

## Category 1: Leadership

### ■ Exceptional / Exceeds Expectations

- *The culture in this building reflects your leadership philosophy in every measurable way.*
- *Your team follows you because they believe in you, not because the org chart says they have to.*
- *You lead from the front even when leading from the front is the harder choice.*
- *The stability you provided during [period/event] was not incidental. It was built deliberately through your daily presence and decision-making.*
- *When the organization needed a model for how to operate under pressure, your building was the model.*
- *You have built something here that will outlast any single performance cycle.*
- *Your ability to see the moves on the chessboard before others do is what separates you from your peers at this level.*
- *The associate engagement results from this period are a direct reflection of the environment you created.*

### ■ Meets Expectations

- *You have demonstrated the ability to influence your team through both your words and your actions.*
- *Your team is comfortable approaching you with opposing ideas, and that trust is only built by a leader who invites it.*
- *You empower the leaders below you by providing timely feedback while giving the right amount of autonomy based on the individual.*
- *You lead effectively within your four walls. The development opportunity is to bring that same presence to settings outside your building.*
- *When nudged, you set a strong example for your peers. The next step is to do this without being nudged.*
- *You treat your associates fairly and take active steps to improve their work experience, and this shows in the results.*

## ■ Improvement Needed / Developing

- *Leadership at this level requires consistency, not just on the days when things are going well.*
- *The team responds to you, but has not yet fully formed around a clear direction set by you.*
- *Effective leadership means setting the standard before the team misses it, not correcting after the fact.*
- *Your associates trust you as a person. The opportunity is to translate that trust into higher performance expectations.*
- *The behaviors you model set the ceiling for what your team believes is possible. Raise the ceiling.*

## Category 2: Accountability

## ■ Exceptional / Exceeds Expectations

- *You own every aspect of this operation: the successes, the gaps, and everything in between.*
- *When something goes wrong on your shift, your first question is always: what do I change? That orientation is the foundation of real accountability.*
- *Your response to actionable feedback is immediate and methodical. You create solutions rather than explanations.*
- *You are widely respected by your peers not because of your title, but because of how consistently you operate with integrity.*
- *The performance of your team is the clearest evidence of how quickly you act when expectations are not being met.*

## ■ Meets Expectations

- *You take ownership of your responsibilities and the results they produce, including when those results fall short.*
- *You produce work that is consistent, timely, and reflective of genuine pride in the role.*
- *You solicit feedback and are viewed as credible among your peers, both signs of real accountability in action.*
- *You demonstrate ethical behavior and operate with a transparency that builds trust across the organization.*

## ■ Improvement Needed / Developing

- *Accountability at this level means owning the environment your team operates in, not just your individual contributions.*
- *When performance gaps occur, the response has been to explain circumstances rather than own outcomes. The expectation going forward is ownership.*

• *Accountability is demonstrated through action, not intention. The coming year requires a shift from one to the other.*

• *Performance in this area has not consistently met the expectations established at the start of the review period.*

• *The standard is clear. The path to meeting it is also clear. What is needed now is consistent execution.*

## Category 3: Excellence and Quality of Work

### ■ Exceptional / Exceeds Expectations

• *In every area of performance, this facility is operating at an above-average or greater level.*

• *The quality of work produced here consistently exceeds what is expected at this point in the operational cycle.*

• *Errors are not tolerated on this team, not because the supervisor demands it, but because the team believes quality is a reflection of who they are.*

• *Your same-day service percentage this year has been among the highest in the network, the result of daily, deliberate operational decisions.*

• *You anticipate where quality will be challenged before the challenge materializes and address it proactively.*

### ■ Meets Expectations

• *Work produced is accurate and delivered on time with a consistency that reflects strong operational discipline.*

• *Quality metrics are within an acceptable range and reflect a team that understands the standard.*

• *You exercise good judgment and operate with an appropriate sense of urgency in the day-to-day.*

• *The customer-facing results of this building's work reflect a team that understands who they are serving.*

### ■ Improvement Needed / Developing

• *Quality and error rates in [specific area] remain above acceptable levels and must improve in the coming cycle.*

• *Work produced is not yet at the level of consistency this role requires. The gap is named specifically: [insert specific metric or behavior].*

• *Excellence is not occasional performance at a high level. It is consistent performance at the expected standard. That consistency is the opportunity here.*

• *The expectation was [specific standard]. The result was [specific result]. That gap is the focus of the coming year.*

## Category 4: Communication

### ■ Exceptional / Exceeds Expectations

- *You communicate with a level of clarity and specificity that raises the standard for those around you.*
- *You have done a masterful job building working relationships with your peer group, accomplished by articulately expressing your thoughts and ideas in both writing and conversation.*
- *Your willingness to listen as often as you speak, rare among strong-minded leaders, is what makes you a trusted voice in the group.*
- *You communicate expectations to your team with a directness that removes ambiguity and enables performance.*

### ■ Meets Expectations

- *You communicate clearly with your team and share information proactively, which keeps your group oriented and aligned.*
- *Written communications are organized and accurate. Verbal communication is direct and appropriate for the audience.*
- *You seek constructive feedback and share information openly with your peers and your team.*
- *You have built cooperative relationships with your colleagues and are an active listener in group settings.*

### ■ Improvement Needed / Developing

- *Communication clarity in [written/verbal/upward/downward] settings has not consistently met the standard expected at this level.*
- *Expectations communicated to the team have not always been specific enough to enable consistent execution. The coming year requires more precision in how direction is delivered.*
- *Your peers value your perspective. The development opportunity is to share it more proactively, without waiting to be asked.*
- *Communication at the next level of leadership requires more than clarity. It requires deliberate influence. That is the skill to build.*

## Category 5: Performance and Productivity

### ■ Exceptional / Exceeds Expectations

- *Productivity, service, housekeeping, and safety are all exemplary and all reflect the standard this leader has set.*
- *You are masterful at seeing the moves on the chessboard. You proactively remove obstacles and know when to take calculated risks.*
- *The trajectory of performance in this building since you took the role is the clearest evidence of your impact as a leader.*

**Kevin Baker**

- *Your team operates with an autonomy and confidence that only comes from having a leader who trusts them and holds them accountable in equal measure.*

### ■ Meets Expectations

- *Performance consistently met expectations in all essential areas of responsibility during this review period.*
- *Productivity metrics are on track and reflect a team that understands the operational expectations of the role.*
- *You have found a reliable routine to the day-to-day that allows you to prioritize effectively and inspect performance in real time.*
- *Safety, service, and quality are all at an acceptable level. The opportunity is to move from acceptable to excellent.*

### ■ Improvement Needed / Developing

- *Performance metrics in [specific area] are below the standard expected and have not improved at the rate required.*
- *The standard for next year is clear: [specific metric]. For the last several months, this standard has not been met with regularity.*
- *As the leader, you must set the standard and drive performance expectations. Make the job easier for your group by demanding process improvements and holding underperformers accountable.*
- *Effort is not the measure. Results are the measure. The results in this area must improve.*

## Category 6: Team Development and Succession

### ■ Exceptional / Exceeds Expectations

- *You have two solid leaders at the next level because of the investment you have made in their development, not because they were talented when they arrived.*
- *Your team started fractured and became, through your leadership, a group full of pride and a sense of shared purpose.*
- *The bench strength in this building reflects a manager who thinks about succession as an ongoing responsibility, not a year-end exercise.*
- *You have hired and developed talented leaders in the network this year. That is the most important thing a manager at your level can do.*

## ■ Meets Expectations

- *You have active Individual Development Plans for your supervisors and are tracking progress against them.*
- *You provide explicit direction and timely feedback to your direct reports, which is enabling their development.*
- *You are growing the next generation of leaders in this building by giving them the right amount of autonomy and the right amount of challenge.*
- *The team understands the succession expectations and is working toward them with clarity and commitment.*

## ■ Improvement Needed / Developing

- *Bench development has not progressed at the pace required. At the current rate, key roles in this building are at risk of having no ready successor.*
- *Development of direct reports requires more intentionality. An IDP that is written but not referenced is not a development plan. It is documentation.*
- *The opportunity is to spend less time on tasks your team should own and more time developing the people who should own them.*
- *Succession readiness in this building must be a stated priority, not an assumed outcome.*

## Category 7: Safety

## ■ Exceptional / Exceeds Expectations

- *Zero OSHA recordable incidents for the full year, an achievement that reflects a leader and team who have made safety a value, not a compliance requirement.*
- *Your team operates with a safety consciousness that is visible the moment you walk into the building.*
- *Safety violations found during facility walks are corrected same-day, communicated to the full team, and documented. That is the standard.*
- *You have built a culture where associates look out for each other's safety because they want to, not because the org chart says they have to.*

## ■ Meets Expectations

- *Safety metrics are within an acceptable range. The expectation is to sustain this performance and build on it.*
- *Weekly safety walks are conducted and documented. Corrective actions are taken and communicated.*
- *Your team understands the safety standard and generally operates within it. The opportunity is to make safety a reflex, not a checklist.*

### ■ Improvement Needed / Developing

• *Safety performance this year ([specific number] recordable incidents) is above acceptable levels and must improve.*

• *The safety program requires more consistent reinforcement at the supervisor level. Safety cannot be a priority one week and an afterthought the next.*

• *Near-miss events that are not documented and addressed are accidents waiting to happen. Reporting culture must improve.*

## Category 8: Closing Statements — How to End a Review

### ■ Closing for a Top Performer

• *Continue to be the rock others feed off of and gravitate toward. You have the wisdom and the leadership command to carry it.*

• *The work you have done this year has been exceptional. What comes next is yours to define, and I believe you are ready to define it.*

• *Great first half. Let's finish well. And then let's build something even better next year.*

• *I am genuinely excited about what this team will accomplish in the year ahead under your leadership.*

• *You start this year with everything you built last year behind you. Lead from the front.*

### ■ Closing for a Solid Contributor

• *You have built a strong foundation this year. The coming year is about building on it, with more intentionality and more ambition.*

• *You are reliable, credible, and valued. The next step is to take on more and to let the team see more of what you are capable of.*

• *This year demonstrated what you can do when you trust your instincts. Trust them more in the year ahead.*

• *The team counts on you. Continue to deserve that trust, and look for ways to expand what it enables you to do.*

### ■ Closing for an Improving Employee

• *The second half of this year showed what you are capable of when you hit your stride. The coming year is about sustaining that and building beyond it.*

• *You took the feedback from last year and did something with it. That is not easy. It is worth naming. Now go further.*

• *The trajectory matters as much as the position. You are moving in the right direction. Keep the momentum.*

- *Growth is not always linear. What matters is that you kept moving forward when it would have been easier not to.*

### ■ Closing for a Struggling Employee

- *This review is written with directness because directness is the most respectful thing I can offer. The gaps are real. They are also fixable. I am invested in seeing that happen.*
- *The coming year is a critical one. The expectation is clear. The support is available. The accountability belongs to you.*
- *I believe you are capable of meeting this standard. What is needed now is consistent execution. Let's get there.*
- *The path forward is clear. Walk it.*

## Category 9: Self-Review Phrases for the Employee

Use these as starting points when completing your own self-review. Customize each with your specific results and evidence.

### ■ Opening Your Self-Review

- *This year was defined by [theme], and the evidence of that is visible in [specific results].*
- *My most significant contribution this year was [accomplishment], achieved through [specific approach].*
- *In the areas where I performed at my best, the defining factor was [behavior or decision].*

### ■ Claiming Accomplishments

- *I led [specific project/initiative] from start to finish, resulting in [specific outcome with metrics].*
- *I identified [specific problem] before it escalated and resolved it by [specific action], preventing [specific impact].*
- *I completed [training/certification/development activity] and applied it directly to [specific work example].*
- *A contribution that may not have been fully visible: I [specific work] which produced [specific result] without requiring escalation.*

## ■ Naming Development Areas Honestly

- *The area where I have the most opportunity to grow is [specific area]. My plan to address this is [specific action with date].*
- *I know I have been slower to [specific behavior] than the role requires. I am committed to [specific change].*
- *I want to develop more confidence in [specific setting or skill]. I am committing to [specific activity] to begin building it.*

## ■ Closing Your Self-Review

- *In the coming year, I want to [specific aspiration]. I believe this is achievable with [specific support or opportunity].*
- *I am ready to take on [specific responsibility]. I would like to discuss what that opportunity might look like.*
- *My goal for next year is to demonstrate [specific competency] at a level that clearly reflects readiness for [specific next step].*

> *These phrases are tools, not scripts. The most powerful self-review language is the language that sounds like you: specific to your work, grounded in your evidence, and honest about both your strengths and your growth. Use these as scaffolding. Then fill them with what actually happened.*

compliance